Math Magic
Slick Tricks with Numbers

by Margaret Thomas

illustrated by Rex Schneider

cover design by Annette Hollister-Papp

Published by Instructional Fair • TS Denison
an imprint of

**McGraw-Hill
Children's Publishing**
*A Division of The **McGraw·Hill** Companies*

Published by Instructional Fair • TS Denison
An imprint of McGraw-Hill Children's Publishing
Copyright © 1998 McGraw-Hill Children's Publishing

Instructional Fair • TS Denison grants permission to the individual purchaser to reproduce pages 7-16, 22-28, 38-48, 56-63, 69-73, and 80-88 in this book for noncommerical, individual, or classroom use only. Reproduction of these pages for an entire school system is strictly prohibited. No other part of this publication may be reproduced, stored in a retrieval system, or transmitted, in any form or by any means, electronic, mechanical, photocopying, recording, or otherwise, without the prior written permission of the publisher.

Send all inquiries to:
McGraw-Hill Children's Publishing
3195 Wilson Drive NW
Grand Rapids, Michigan 49544

All Rights Reserved • Printed in the United States of America

Math Magic: Slick Tricks with Numbers—grades 5–8
ISBN: 1-56822-617-9

2 3 4 5 6 7 8 9 PHXBK 07 06 05 04 03

Table of Contents

Magical Mathematical Patterns
Teacher Notes 1
Switch-a-Roo 7
Magical Nines 8
Could You Repeat That? 9
Are All Things Equal? 10
Geometric Numbers 11
"Sum" Fun with Fibonacci—I 13
"Sum" Fun with Fibonacci—II 14
"Sum" Products 15
Was Fibonacci a Square? 16

Magical Computational Tricks and Shortcuts
Teacher Notes 17
Chet: The Sum Checker 22
Chet: The Product Checker 23
Multiplication Madness 24
Multiplication Magic 25
Win Some, Lose Some Multiplication . . 26
Pascal's Triangle 27
Integer Trees 28

Magical Predictions
Teacher Notes 29
Back to the Beginning 38
Age Predictor/Birthday Predictor 40
Pick a Number—Any Number/Calculator Grid 41
Double Vision/Multi-vision 42
Caught in the Middle/Digit Predictor . . . 43
Sum Predictor 44
Human Calculator 45
Fantastic Fractions 46
High Noon 47
A Mouse in the House 48

Magic Squares, Triangles, and Circles
Teacher Notes 49
Magic Squares 56
Fraction Magic Squares 57
Decimal Magic Squares 58
Equation Magic Square 59
A Magical Magic Square 60
Magic Square Triangle 61
Magic Triangles 62
Magic Circles 63

Math Magic with Cards and Cups
Teacher Notes 64
Magic Card/Pick a Card—Any Card . . . 69
Magic Deal/Card Flip 70
Card Predictor/Three Card Guess 71
Mind Reading Cards 72
Four Cups/Three Cups 73

Seasonal Magic Activities
Teacher Notes 74
Calendar Squares 80
Columbus Day Magic 81
November—Ninth Month? 82
November Now! 83
Twelve Days of Christmas 84
Happy Birthday, Mr. President! 85
A Mathematical Valentine 86
Shamrock Story 87
April Showers, May Flowers 88

Answer Key 89

Introduction

Math Magic: Slick Tricks with Numbers provides the teacher with magic tricks, activities, and shortcuts that are mathematically based. Concepts such as even/odd, place value, factors and multiples, fraction-decimal equivalents, casting out nines, algebraic identities, and integer trees are used to explain the tricks and develop the computational shortcuts. Activities explore and explain the Fibonacci series, geometric numbers, and Pascal's triangle. The activities may be used for class-wide discussions as well as partner and small group settings. Several activities ask students to make predictions and describe patterns. Student responses can be written or oral. Communication of mathematical concepts among students and between the teacher and students is emphasized.

The book includes six sections: Magical Mathematical Patterns; Magical Computational Tricks and Shortcuts; Magical Predictions; Magic Squares, Triangles, and Circles; Math Magic with Cards and Cups; and Seasonal Magic Activities. An Answer Key is provided. Although several activities complement one another, the order of presentation is left to the discretion of the teacher as each activity can be presented alone.

Each section contains teacher notes and blackline masters of the activities. The teacher notes contain a brief description of the activity, a list of needed materials, and comments that include an explanation of the activity, mathematical proofs, extensions, and suggestions for additional activities and class discussions. The blackline masters of the activities may be used to make a transparency for class-wide discussions or to make copies for students to use individually or in small-group settings. When two activities are printed on a page, it is recommended that the teacher make a copy of the page, make the necessary transparency and copies, and then cut the activities apart.

Throughout the materials, mathematics is used to explain the mystery of an activity and to let students be excited by, and explore, its magic.

Teacher Notes

The activities in this section have students explore and describe mathematical patterns based on place value, the number 9, fraction/decimal equivalents, geometric numbers, and the Fibonacci series. "Switch-a-Roo" and "Are All Things Equal?" provide some surprising results that can be explained mathematically.

Switch-a-Roo

Activity Students compute products of pairs of numbers that contain the same four digits and are asked to determine the restrictions on the factors for the products to be equal.

Comments The comparison of the factors used in the multiplication problems requires considering place value.

The factors in each pair of problems use the same digits so that the products of the tens place and the ones place digits are equal.

For example, consider   39   93
                       x 62   and   x 26

                   3 x 6 = 18   equals   9 x 2 = 18

The product of the tens digits equals the product of the ones digits.

In general, consider the pair of problems:
    10A+B   10B+A   where **AC = BD**
    x 10C+D   and   x 10D+C

For the first problem, (10A + B) x (10C + D) = 100 AC + 10AD + 10BC + BD
Since the digits were selected so that **AC = BD**,
100AC + 10AD + 10BC + **BD** = 100BD + 10AD + 10BC + AC, which equals
(10B + A)(10D + C), the second problem.

© Instructional Fair • TS Denison   1   IF2506 Math Magic: Slick Tricks with Numbers

Magical Nines

Activity Students complete three patterns based on the number 9 and are then asked to describe the patterns.

Comments The activity is appropriate for small-group work with calculators.
1. 8 x 9 = 8 x 9(1) = 72
   8 x 99 = (8 x 9)(10 + 1) = 720 + 72 = 792
   8 x 999 = (8 x 9)(100 + 10 + 1) = 7,200 + 720 + 72 = 7,992
   Largest place value is 7. Ones digit is 2. Other digits are 7 + 2 or 9.

2. 99 x 12 = (100 - 1) x 12 = 100 x 12 - 12
   In general, 99 x n = (100 - 1) x n = 100 x n - n

3. 1 x 9,109 = 9,109
   2 x 9,109 = 18,218
   In general, n x 9,109 = (n x 9,000) + (n x 100) + (n x 0) + (n x 9) = 9,000n + 100n + 9n.
   The digits in the product are n x 9 followed by n followed by n x 9.

Could You Repeat That?

Activity Students explore the pattern of the digits in the repeating decimal equivalents for given fractions.

Comments This activity reinforces fraction and decimal equivalences. A topic for class discussion would be how to tell if a fraction has a terminating or repeating decimal equivalent—a fraction with a denominator containing prime factors of only 2s and/or 5s will terminate since the only prime factors of 10 are 2 and 5. All other fractions repeat. A repeating decimal will repeat in no more decimal places than one less than the equivalent fraction's denominator.

The most interesting pattern is perhaps the ⅐s. The decimal equivalent of ⅐ repeats in six digits: ⅐ = 0.142857142857 . . . When the decimal 0.142857 is multiplied by the numbers 1 through 6, the products have the same order of digits as the original decimal.

Are All Things Equal?

Activity Students complete fraction problems which seem to prove that multiplication and division are the same as addition and subtraction.

Comments It must be noted that the equality statements rely on the pattern of the numbers used. The proofs offer a review of algebra concepts such as operating with rational expressions, factoring trinomials, and simplifying expressions.

Multiplication	Addition		Multiplication	Subtraction
$1\frac{1}{a} \times (a+1)$	? $1\frac{1}{a} + (a+1)$		$a \times \dfrac{a}{a+1}$	? $a - \dfrac{a}{a+1}$
$\dfrac{a+1}{a} \times \dfrac{a+1}{1}$	$2 + \frac{1}{a} + a$		$\dfrac{a^2}{a+1}$	$\dfrac{a^2 + a - a}{a+1}$
$\dfrac{a^2 + 2a + 1}{a}$			$\dfrac{a^2}{a+1}$ =	$\dfrac{a^2}{a+1}$
$a + 2 + \frac{1}{a}$	= $a + 2 + \frac{1}{a}$			

Division			Addition
$\left(a + \dfrac{1}{a+2}\right) \div \dfrac{a+1}{a+2}$	?		$a + \dfrac{1}{a+2} + \dfrac{a+1}{a+2}$
$\dfrac{a^2 + 2a + 1}{a+2} \times \dfrac{a+2}{a+1}$			$\dfrac{a^2 + 2a + 1 + a + 1}{a+2}$
$\dfrac{(a+1)^2}{a+2} \times \dfrac{a+2}{a+1}$			$\dfrac{a^2 + 3a + 2}{a+2}$
$a + 1$			$\dfrac{(a+2)(a+1)}{a+2}$
$a + 1$	=		$a + 1$

Division			Subtraction
$\left(a + \dfrac{1}{a-2}\right) \div (a-1)$	?		$a + \dfrac{1}{a-2} - (a-1)$
$\dfrac{a^2 - 2a + 1}{a-2} \times \dfrac{1}{a-1}$			$\dfrac{a^2 - 2a + 1 - a^2 + 3a - 2}{a-2}$
$\dfrac{(a-1)^2}{a-2} \times \dfrac{1}{a-1}$			$\dfrac{a-1}{a-2}$
$\dfrac{a-1}{a-2}$	=		$\dfrac{a-1}{a-2}$

Geometric Numbers

Activity Students explore series of geometric numbers: triangular, square, pentagonal, and hexagonal from a geometric viewpoint and a pattern-based viewpoint. Additional series are listed by finding sums and differences of the geometric numbers. Students are asked to describe the various series.

Comments TRIANGULAR NUMBERS: 1, 3, 6, 10, 15, 21, 28, 36, 45, 55, . . .
The differences: 2, 3, 4, 5, 6, 7, 8, 9, 10, . . . form the series of consecutive integers greater than 1, or $1 + 1n$ (n is the number of the term). The sums: 4, 9, 16, 25, 36, 49, 64, 81, 100 form the series of squares 4 and greater.

SQUARE NUMBERS: 1, 4, 9, 16, 25, 36, 49, 64, 81, 100, . . .
The differences: 3, 5, 7, 9, 11, 13, 15, 17, 19, . . . form the series of odds greater than 1 or $1 + 2n$ (n is the number of the term).

PENTAGONAL NUMBERS: 1, 5, 12, 22, 35, 51, 70, 92, 117, 145, . . .
The differences: 4, 7, 10, 13, 16, 19, 22, 25, 28, . . . form the series of $1 + 3n$ (n is the number of the term).

HEXAGONAL NUMBERS: 1, 6, 15, 28, 45, 66, 91, 120, 153, 190, . . .
The differences: 5, 9, 13, 17, 21, 25, 29, 33, 37, . . . form the series of $1 + 4n$ (n is the number of the term).

"Sum" Fun With Fibonacci—I

Activity Students discover that the sum of 10 consecutive numbers of the Fibonacci series equals 11 times the seventh number.

Comments Let A stand for the first number and B stand for the second number of ten consecutive elements of the Fibonacci series or any Fibonacci-like series. The ten consecutive elements are

$$A$$
$$B$$
$$A + B$$
$$A + 2B$$
$$2A + 3B$$
$$3A + 5B$$
$$5A + 8B$$
$$8A + 13B$$
$$13A + 21B$$
$$21A + 34B$$

The sum of the ten elements is $55A + 88B$. The quotient of $(55A + 88B)$ and 11 is $5A + 8B$ which is the seventh element of the series.

"Sum" Fun With Fibonacci—II

Activity Students predict the sum of a set of numbers of the Fibonacci series.

Comments Let A stand for the first number and B stand for the second number of the Fibonacci series. Suppose the line is drawn under the fifth number of the series.

$$A$$
$$B$$
$$A + B$$
$$A + 2B$$
$$\underline{+2A + 3B}$$

Sum of numbers above the line: $5A + 7B$

The seventh number of the series would be $5A + 8B$.
$5A + 8B = 5A + 7B + B = 5A + 7B + 1$, since $B = 1$.
$5A + 7B = 5A + 7B + (1 - 1) = 5A + (7B + 1) - 1 = 5A + 8B - 1$.
Sum above the line ($5A + 7B$) is 1 less than the number 2 places below the line ($5A + 8B - 1$). The argument can be extended to other numbers.

"Sum" Products

Activity Students find products of consecutive terms from the Fibonacci series and compare them to the products of preceding and following terms. Then they compare the square of a term to the product of its preceding and following terms and describe the patterns.

Comments The two products differ by one. However, there is a pattern to which of the two products is greater. Number the terms and consider the following:

1	1	2	3	5	8	13	21	34	55	89	144	233...
n=1	2	3	4	5	6	7	8	9	10	11	12	13...

The product of two consecutive terms: $T_n \times T_{n+1} = T_{n-1} \times T_{n+2} \pm 1$
Case 1: If n is even, then it is - 1.
Ex.: 3 is the fourth term, so n is even. $3 \times 5 = 2 \times 8 - 1$ or $15 = 16 - 1$

Case 2: If n is odd, then it is + 1.
Ex.: 2 is the third term, so n is odd. $2 \times 3 = 1 \times 5 + 1$ or $6 = 5 + 1$

The square of a term: $T_n^2 = T_{n-1} \times T_{n+1} \pm 1$
Case 1: If n is even, then it is - 1.
Ex.: 3 is the fourth term, so n is even. $3^2 = 2 \times 5 - 1$ or $9 = 10 - 1$

Case 2: If n is odd, then it is + 1.
Ex.: 2 is the third term, so n is odd. $2^2 = 1 \times 3 + 1$ or $4 = 3 + 1$

Was Fibonacci a Square?

Activity Students form series by adding the squares of the terms of the Fibonacci series and calculating the differences of the sums. Students are asked to describe the series they have written.

Comments The series formed by adding the pairs of the squares includes the odd terms of the Fibonacci series starting with 5 (the fifth term).

The differences of pairs of the sums give the even terms of the Fibonacci series starting with 3 (the fourth term).

Name _____

Switch-a-Roo

Compute the following pairs of products. Compare the factors of each pair.

    39   93   48   84
x 62   and   x 26   x 21   and   x 12

    26   62   36   63
x 31   and   x 13   x 21   and   x 12

    39   93   36   63
x 31   and   x 13   x 42   and   x 24

    24   42   12   21
x 84   and   x 48   x 42   and   x 24

Can you determine the rule for the numbers used to obtain the desired results?

_____
_____
_____
_____

Hint: Problems such as 56 and 65 do not have the same results.
                                 x 42   x 24

© Instructional Fair • TS Denison   IF2506 Math Magic: Slick Tricks with Numbers

Name _____

Magical Nines

Complete the patterns:

1. 8 x 9 = 72
   8 x 99 = 792
   8 x 999 = _____
   8 x 9,999 = _____
   8 x 99,999 = _____
   8 x 999,999 = _____
   8 x 9,999,999 = _____

   Describe the pattern. _____

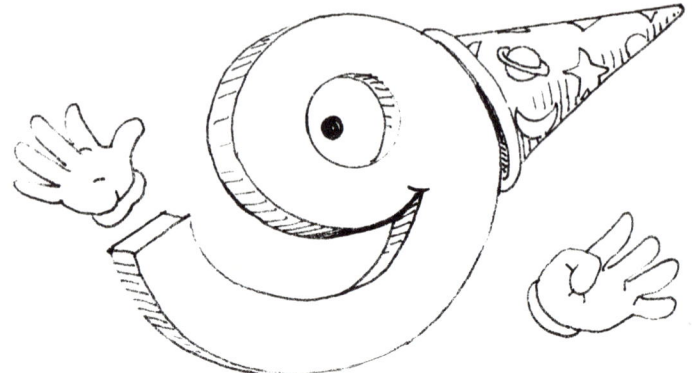

2. 99 x 12 = 1,188   100 x 12 - 12 = 1,188
   99 x 23 = _____   100 x 23 - ____ = _____
   99 x 34 = _____   100 x ____ - ____ = _____
   99 x 45 = _____   100 x ____ - ____ = _____
   99 x 56 = _____   100 x ____ - ____ = _____
   99 x 67 = _____   100 x ____ - ____ = _____
   99 x 78 = _____   100 x ____ - ____ = _____
   99 x 89 = _____   100 x ____ - ____ = _____

   Describe the pattern. _____

3. 1 x 9,109 = 9,109
   2 x 9,109 = _____
   3 x 9,109 = _____
   4 x 9,109 = _____
   5 x 9,109 = _____
   6 x 9,109 = _____
   7 x 9,109 = _____
   8 x 9,109 = _____
   9 x 9,109 = _____

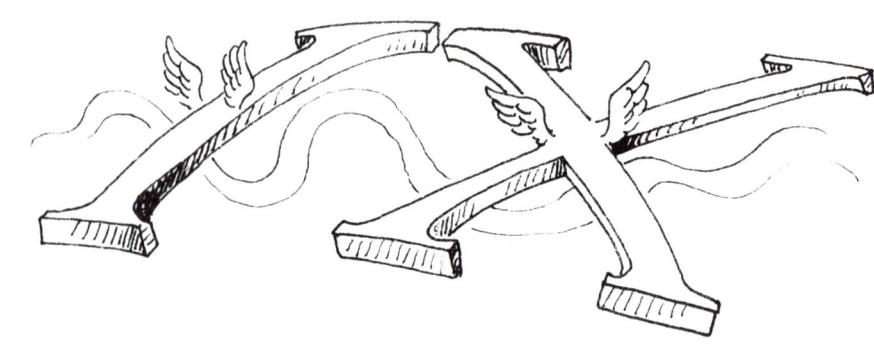

   Describe the pattern. _____

Name _____

Could You Repeat That?

Change the following fractions to their decimal equivalents. Describe the patterns.

1/3 = _____   1/9 = _____   1/11 = _____

2/3 = _____   2/9 = _____   2/11 = _____

3/3 = _____   3/9 = _____   3/11 = _____

                   4/9 = _____   4/11 = _____

1/7 = _____   5/9 = _____   5/11 = _____

2/7 = _____   6/9 = _____   6/11 = _____

3/7 = _____   7/9 = _____   7/11 = _____

4/7 = _____   8/9 = _____   8/11 = _____

5/7 = _____   9/9 = _____   9/11 = _____

6/7 = _____   10/11 = _____

7/7 = _____   11/11 = _____

1. What pattern do you find in the decimal equivalents for the 1/3s and the 1/9s? _____
   _____

2. What pattern do you find in the decimal equivalents for the 1/11s? _____
   _____

3. What pattern do you find in the decimal equivalents for 1/7s? _____
   _____

4. Explain how 1 = 0.9999999.... _____

© Instructional Fair • TS Denison   IF2506 Math Magic: Slick Tricks with Numbers

Name _____

Are All Things Equal?

Everyone knows that multiplication is not the same as addition or subtraction.

    $5 \times 4 \neq 5 + 4$ $5 \times 4 \neq 5 - 4$
       $20 \neq 9$ $20 \neq 1$

But consider the following and complete the patterns:

Multiplication		Addition	Multiplication		Subtraction
$1\frac{1}{2} \times 3$	=	$1\frac{1}{2} + 3 = 4\frac{1}{2}$	$1 \times \frac{1}{2}$	=	$1 - \frac{1}{2} = \frac{1}{2}$
$1\frac{1}{3} \times 4$	=	$1\frac{1}{3} + \underline{\quad} = 5\frac{1}{3}$	$2 \times \frac{2}{3}$	=	$\underline{\quad} - \frac{2}{3} = 1\frac{1}{3}$
$1\frac{1}{4} \times 5$	=	$1\underline{\quad} + \underline{\quad} = 6\frac{1}{4}$	$3 \times \frac{3}{\underline{\quad}}$	=	$3 - \underline{\quad} = 2\frac{1}{4}$
$1\frac{1}{5} \times \underline{\quad}$	=	$1\underline{\quad} + \underline{\quad} = 7\frac{1}{5}$	$4 \times \frac{4}{\underline{\quad}}$	=	$\underline{\quad} - \underline{\quad} = 3\frac{1}{5}$
$1\underline{\quad} \times \underline{\quad}$	=	$1\underline{\quad} + \underline{\quad} = 8\frac{1}{6}$	$5 \times \underline{\quad}$	=	$\underline{\quad} - \underline{\quad} = 4\frac{1}{6}$

Everyone knows that division is not the same as addition or subtraction.

     $20 \div 4 \neq 20 + 4$ $20 \div 4 \neq 20 - 4$
       $5 \neq 24$ $5 \neq 16$

But consider the following and complete the patterns:

Division		Addition	Division		Subtraction
$1\frac{1}{3} \div \frac{2}{3}$	=	$1\frac{1}{3} + \frac{2}{3} = 2$	$4\frac{1}{2} \div 3$	=	$4\frac{1}{2} - 3 = 1\frac{1}{2}$
$2\frac{1}{4} \div \frac{3}{4}$	=	$2\frac{1}{4} + \underline{\quad} = 3$	$5\frac{1}{3} \div 4$	=	$5\frac{1}{3} - \underline{\quad} = 1\frac{1}{3}$
$3\frac{1}{5} \div \frac{4}{5}$	=	$3\underline{\quad} + \underline{\quad} = 4$	$6\frac{1}{4} \div 5$	=	$6\underline{\quad} - \underline{\quad} = 1\frac{1}{4}$
$4\frac{1}{6} \div \underline{\quad}$	=	$4\underline{\quad} + \underline{\quad} = 5$	$7\frac{1}{5} \div \underline{\quad}$	=	$7\underline{\quad} - \underline{\quad} = 1\frac{1}{5}$
$5\underline{\quad} \div \underline{\quad}$	=	$5\underline{\quad} + \underline{\quad} = 6$	$8\underline{\quad} \div \underline{\quad}$	=	$8\underline{\quad} - \underline{\quad} = 1\frac{1}{6}$

Write the next entry for each of the four patterns.

    Multiplication   Addition   Multiplication   Subtraction

    Division   Addition   Division   Subtraction

© Instructional Fair • TS Denison   IF2506 Math Magic: Slick Tricks with Numbers

Geometric Numbers

Name _____

A *geometric number series* is the series of numbers indicating the objects needed to construct increasingly larger versions of the indicated geometric shape.

TRIANGULAR NUMBERS—complete the pattern:

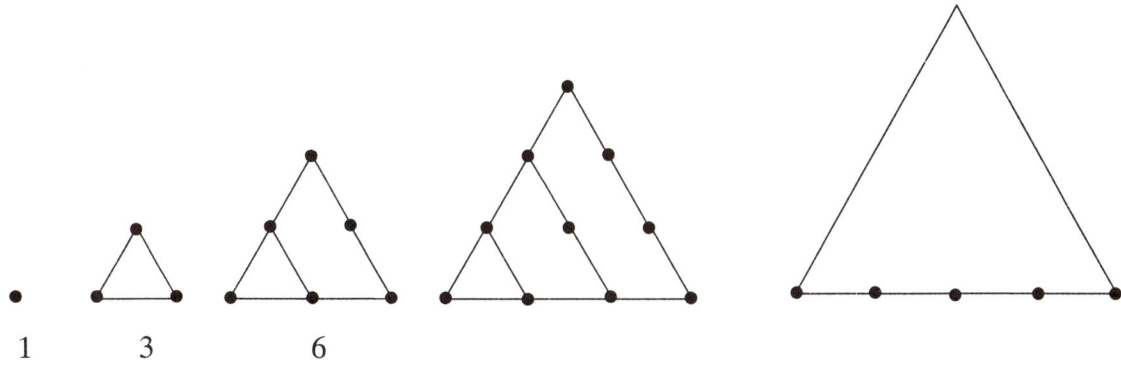

1   3   6

1. List the first ten triangular numbers: _____

2. List the differences between each pair of numbers (2nd - 1st, 3rd - 2nd, 4th - 3rd, etc.):
   _____ Describe this series. _____

3. List the sums of each pair of numbers (1st + 2nd, 2nd + 3rd, 3rd + 4th, etc.):
   _____ Describe this series. _____

SQUARE NUMBERS—complete the pattern:

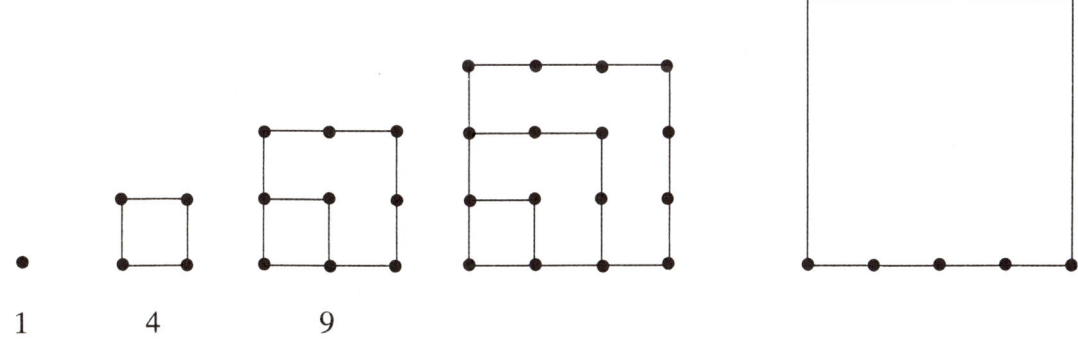

1   4   9

4. List the first ten square numbers: _____

5. List the differences between each pair of numbers (2nd - 1st, 3rd - 2nd, 4th - 3rd, etc.)

   _____ Describe this series. _____

Name _____

PENTAGONAL NUMBERS—complete the pattern:

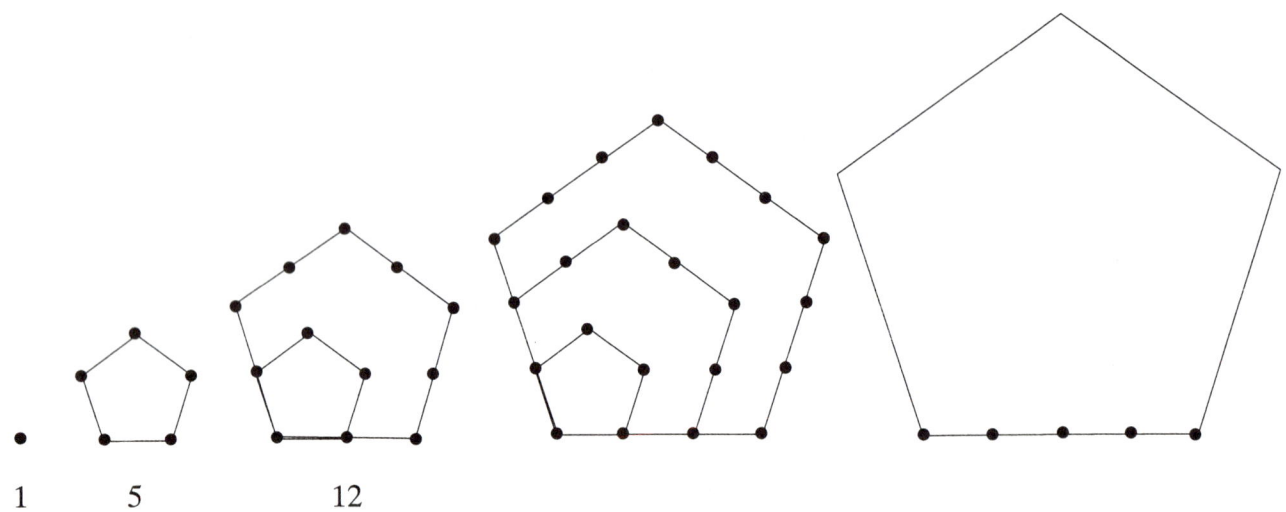

1   5   12

6. List the first ten pentagonal numbers. Hint: Look at the differences between each pair.
_____

7. List the differences between each pair of numbers (2nd - 1st, 3rd - 2nd, 4th - 3rd, etc.):
_____ Describe this series. _____

HEXAGONAL NUMBERS—complete the pattern:

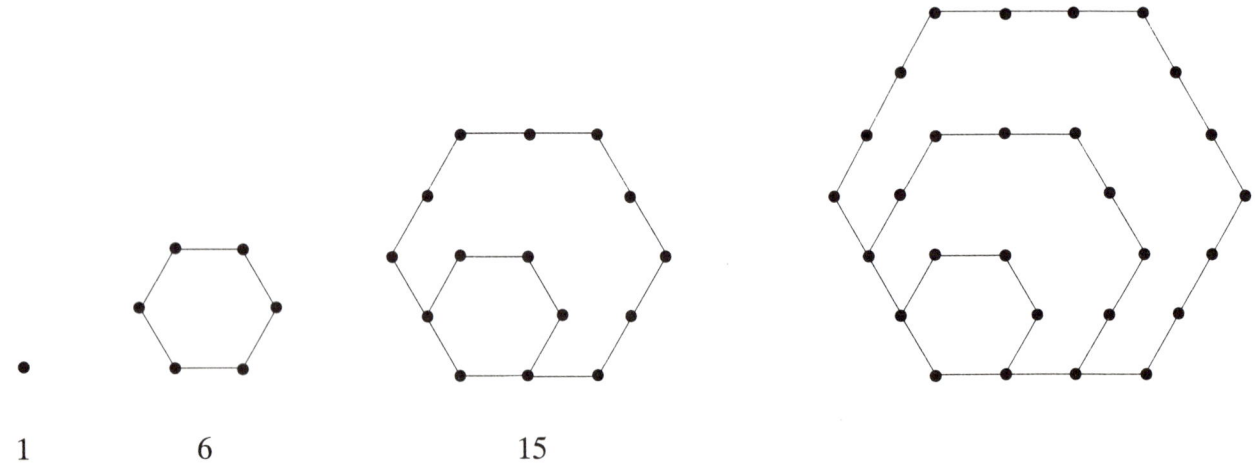

1   6   15

8. List the first ten hexagonal numbers. Hint: Look at the differences between each pair.
_____

9. List the differences between each pair of numbers (2nd - 1st, 3rd - 2nd, 4th - 3rd, etc.):
_____ Describe this series. _____

Name _____

"Sum" Fun With Fibonacci—I

In 1202, Leonardo of Pisa (known as Fibonacci) wrote about the series of numbers given below:

  1   1   2   3   5   8   13   21   34   55   89   144   233   377   610 . . .

Each number of the series, after the second, is the sum of the two preceding numbers.

The numbers of the Fibonacci sequence are found throughout nature from the spirals of a pine cone to the center of a sunflower to the branching of tree limbs.

To discover an interesting feature of the series, do the following:

1. Make a column of any ten consecutive numbers from the series:

2. Compute the sum of the ten numbers. _____

3. Divide the sum by 11. _____

4. Locate the seventh number of the column you used. _____ Surprised?

Try another set of ten consecutive numbers from the series.

This fascinating feature is true for "Fibonacci-like" series also. Follow the steps using this series of numbers: 4, 7, 11, 18, 29, 47, 76, 123, 199, 322. (Note: Each number after the second is the sum of the two preceding numbers.)

Name _____

"Sum" Fun With Fibonacci—II

Amaze your friends with your lightning-fast ability to add a column of numbers.

1. List the terms in the Fibonacci series in a column.

2. Have a friend draw a line under any number (except the last two).

3. Announce the sum of the numbers above the line immediately.

How? The sum of the numbers above the line is 1 less than the second number below the line.

Examples:

Line under 13 → 1 + 1 + 2 + 3 + 5 + 8 + 13 = 33 = 34 - 1

Line under 610 → 1 + 1 + 2 + 3 + 5 + 8 + 13 + 21 + 34 + 55 + 89 + 144 + 233 + 377 + 610 = 1,596 = 1,597 - 1

Try other sums:
    line under 233   line under 1,597

    line under 377   line under 6,765

1
1
2
3
5
8
13
21
34
55
89
144
233
377
610
987
1,597
2,584
4,181
6,765
10,946
17,711

© Instructional Fair • TS Denison   IF2506 Math Magic: Slick Tricks with Numbers

"Sum" Products

Name _____

The Fibonacci series is based on sums. Each term, after the second, is the sum of the two preceding terms.

1   1   2   3   5   8   13   21   34   55   89   144   233 . . .

The series has some interesting **product** number patterns. Complete the following:

1. Choose two consecutive terms from the series (not 1 and 1)   _____ _____

2. Write the preceding term and the following term.   _____ _____

3. Find the product of each pair.   (1) _____   (2) _____

4. Choose two other consecutive terms.   _____ _____

5. Write their preceding and following terms.   _____ _____

6. Find the product of each pair.   (4) _____   (5) _____

7. Choose another pair of terms, find the products, and describe the pattern.

_____

To discover a similar pattern:

8. Choose any term and square it.   _____   _____

9. Find the product of the preceding term and the following term.   _____

10. Choose another term and square it.   _____   _____

11. Find the product of the preceding term and the following term.

_____

12. Describe the pattern of the square of a term, and the product of its preceding and following terms.

_____
_____

$1 + 2 = 3$

$2 + 3 = 5$

$3 + 5 = 8$

Was Fibonacci a Square?

Name _____

The Fibonacci series is a series of numbers in which each term, after the second, is the sum of the two preceding terms.

1   1   2   3   5   8   13   21   34   55   89   144   233 . . .

The squares of the terms provide some interesting number patterns.

Calculate the square of the terms listed.

1^2 2^2 3^2 5^2 8^2 13^2 21^2 34^2 55^2 89^2 144^2 233^2

List the squares: ___ ___ ___ ___ ___ ___ ___ ___ ___ ___ ___ ___

Find the sum of
each pair of squares
(1st + 2nd, 2nd + 3rd, . . .): ___ ___ ___ ___ ___ ___ ___ ___ ___ ___ ___

Find the difference of
each pair of sums:
(2nd - 1st, 3rd - 2nd, . . .): ___ ___ ___ ___ ___ ___ ___ ___ ___ ___

Do the numbers in the last two lists look familiar? How could you describe them?

_____

© Instructional Fair • TS Denison   IF2506 Math Magic: Slick Tricks with Numbers

Magical Computational Tricks and Shortcuts

Teacher Notes

The tricks and shortcuts explained here are based on casting out nines, algebraic identities, Pascal's triangle, and integer trees. Chet Checker and Professor Mad Math provide methods for checking sums and products. Pascal's triangle and integer trees are used to shorten the procedure for multiplying binomials.

Chet: The Sum Checker
Chet: The Product Checker

Activity Students check addition, subtraction, and multiplication problems by adding the digits in a problem until a single digit remains for each number; adding, subtracting, or multiplying the single digits; and comparing the result to the sum of the answer's digits.

Comments The two activities use the *casting out nines* method of checking addition, subtraction, and multiplication problems. The term *casting out nines* comes from counting 9s as 0s and the fact that any number can be written as the sum of 9 times the tens digit and the sum of the digits:

$10T + U = (9T + T) + U = 9T + (T + U)$.

The repeated addition of the digits to a single digit extends the *casting out nines* to the sum of the problem addends, which is the answer. The method applies to subtraction as well. Be certain to subtract the digits representing the problem numbers. If the difference is negative, add 9 before checking:

$$\begin{array}{lll} 157{,}983 \rightarrow 33 \text{ or } 24^* \rightarrow 6 & & \text{*counting 9s as 0s} \\ \underline{-\ 6{,}856} \rightarrow 25 \rightarrow & \underline{-\ 7} & \\ 151{,}127 \rightarrow 17 \rightarrow 8 & -1 + 9 = 8 & \end{array}$$

Since multiplication is repeated addition, the method works with factors and products as long as the digit sums of the factors are multiplied.

$$\begin{array}{lll} 2{,}914 \rightarrow 16 \text{ or } 7^* \rightarrow\ \ 7 & & \text{*counting 9s as 0s} \\ \underline{\times\ 856} \rightarrow 19 \rightarrow & \underline{\times\ 1} & \\ 2{,}494{,}384 \rightarrow 34 \rightarrow 7 & 7 & \end{array}$$

© Instructional Fair • TS Denison   IF2506 Math Magic: Slick Tricks with Numbers

Multiplication Madness

Activity Students find the product of two numbers by repeatedly dividing one number by 2 and multiplying the other number by 2.

Comments Professor Mad Math's method is based on the binary system. Since the factors on the right are repeatedly multiplied by 2, the computation is based on powers of 2. Professor Mad Math's technique for the following problems:
1. Divide the left factor by 2 (drop any remainder).
2. Multiply the right factor by 2.
3. Repeat step 1 and step 2 until you reach 1 on the left.
4. Cross out the rows that have even factors on the left.
5. Add the factors on the right that have not been crossed out.
6. Is the sum the product of the first two numbers?

Consider the examples: Remember that the professor multiplies the right side factors by two:

~~20~~	x	~~34~~	The 34 is treated like	~~2^0~~	x	~~34 = 34~~
~~10~~	x	~~68~~	The 68 is treated like	~~2^1~~	x	~~34 = 68~~
5	x	136	The 136 is treated like	2^2	x	34 = 136
~~2~~	x	~~272~~	The 272 is treated like	~~2^3~~	x	~~34 = 272~~
1	x	544	The 544 is treated like	2^4	x	34 = 544
		680				

Rows 2^0, 2^1, and 2^3 are eliminated.
Rows 2^2 and 2^4 are added together, which is equivalent to
$(4 \times 34) + (16 \times 34) = (4 + 16) \times 34 = 20 \times 34 = 680$.
The number 20 in the binary system is 10100 or $2^4 + 2^2 = 16 + 4$.

75	x	13		2^0	x	13 = 13
37	x	26		2^1	x	13 = 26
~~18~~	x	~~52~~		~~2^2~~	x	~~13 = 52~~
9	x	104		2^3	x	13 = 104
~~4~~	x	~~208~~		~~2^4~~	x	~~13 = 208~~
~~2~~	x	~~416~~		~~2^5~~	x	~~13 = 416~~
1	x	832		2^6	x	13 = 832
		975				

Rows 2^2, 2^4, and 2^5 are eliminated.
Rows 2^0, 2^1, 2^3, and 2^6 are added together which is equivalent to
$(1 \times 13) + (2 \times 13) + (8 \times 13) + (64 \times 13) = (1 + 2 + 8 + 64) \times 13 = 75 \times 13 = 975$.
The number 75 in the binary system is 1001011 or
$2^6 + 2^3 + 2^1 + 2^0 = 64 + 8 + 2 + 1$.

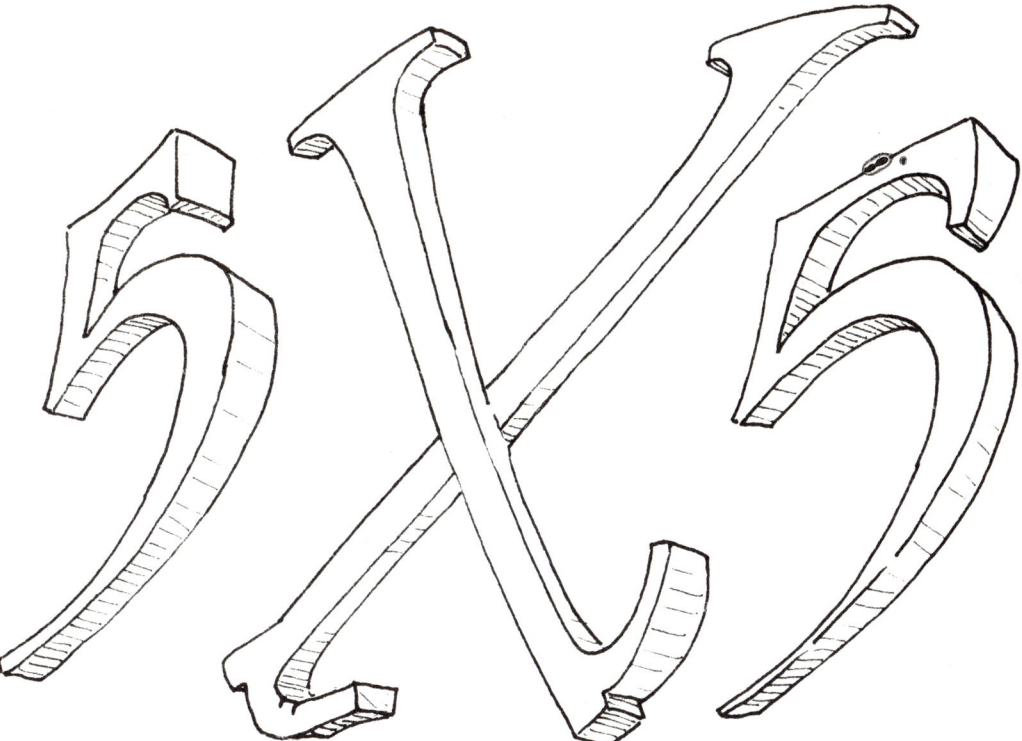

Multiplication Magic

Activity Students discover the pattern of squaring any number ending in 5.

Comments The number of hundreds is the product of the tens digit and one more than the tens digit. Any two-digit number ending in 5 can be written as 10T + 5.
$$(10T + 5)(10T + 5) = 100T^2 + 100T + 25$$
$$= 100T(T + 1) + 25$$

The pattern can be extended to three-digit numbers ending in 5 since T in the above example is not restricted to a single digit.

Win Some, Lose Some Multiplication

Activity Students complete multiplication problems to discover that the product of the sum and difference of two numbers equals the difference of their squares.

Comments The TRICK is based on the algebra fact that when one multiplies the sum and difference of two numbers, the result is the difference of the squares of the two numbers.

$$(A + B)(A - B) = A^2 - 2AB + 2AB - B^2 = A^2 - B^2$$
$$33 \times 27 = (30 + 3)(30 - 3) = 900 - 9 = 891$$

Pascal's Triangle

Activity   Students use Pascal's triangle to calculate the coefficients of binomial expansions of $(x + y)^n$.

Comments   This activity sheet presents Pascal's triangle and several examples of its use in calculating the coefficients of binomial expansions. Problems include completing rows of the triangle, completing binomial expansions by finding coefficients, and completing an expansion by filling in coefficients and exponents.

Pascal's triangle contains several patterns. Some patterns include
- each term is the sum of the two terms diagonally above it
- the first diagonal is all ones
- the second diagonal contains consecutive positive integers
- the third diagonal is a series of triangular numbers

Additional activities could include having students discuss and write the various patterns they discover. Exploring Pascal's triangle is appropriate for small- and large-group work.

Integer Trees

Activity Students complete integer trees to find the product of two and three binomials.

Comments This activity is appropriate for students with experience in multiplying binomials. It provides a method for determining the coefficients of the product of several binomials. The activity sheet gives three examples of using integer trees to write the product of a constant by a binomial, and the product of two and three binomials. Two problems are given with partial integer trees for students to complete. A third problem has students write a complete integer tree.

For example:
$(4x + 1)(2x - 5)(3x + 7) = 24x^3 - 10x^2 - 139x + 35$

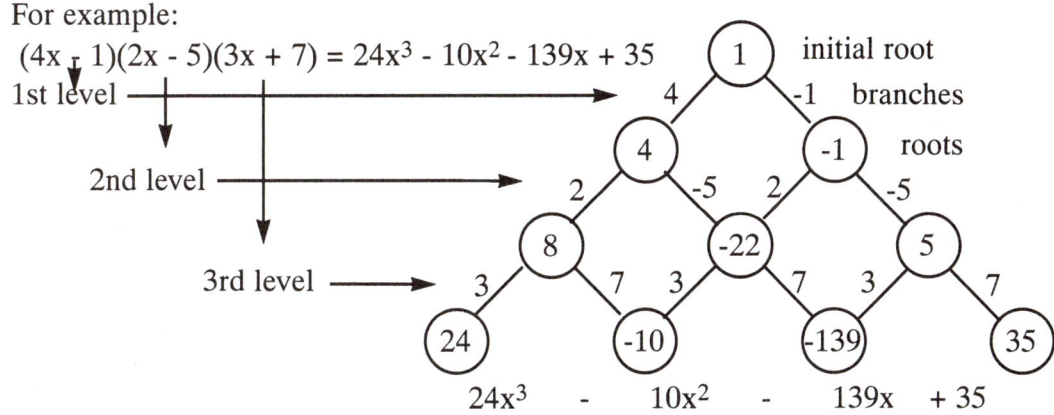

The initial root is 1 unless a constant is multiplying the binomial as in $5(3x + 4)$. Then, the constant is the initial root. In the first level, the left branch is the coefficient of x and the right branch is the constant of the first binomial. The roots of the first level are found by multiplying the initial root by the branches.

The second and subsequent levels are written using the coefficient of x and the constant of the subsequent binomials as the left and right branches. "Middle roots" are the sum of the products of the attached branches.

Name _____

Chet: The Sum Checker

Chet Checker claims to be the fastest "answer checker" in the world. Chet adds the digits of the numbers in a problem repeatedly until a single digit remains for each number. To save time, he counts 9s as 0s. He then adds the digits in the given answer the same way. If the sum of the digits in the problem equals the sum of the digits in the answer, the problem receives Chet Checker's OK.

Example:

```
       4,370  →  14  →   5              14,964  →  15  →   6
         739  →  10  →   1              13,579  →  16  →   7
       + 274  →  13  →  +4             + 4,321  →  10  →  +1
       5,383           10   → 1         31,864           14  → 5
5 + 3 + 8 + 3  →  19  →  1  OK    3 + 1 + 8 + 6 + 4  →  22  →  4   NO
```

Which problems would receive Chet Checker's OK?

1. 2,567 →	2. 953 →	3. 8,932 →	4. 765 →
3,211 →	654 →	1,164 →	298 →
+ 5,982 →	+ 2,789 →	+ 7,623 →	+ 562 →
11,760	4,296	17,619	1,625

Chet Checker uses the method for subtraction problems also. If the difference of the problem digit sums is negative, he adds 9 before checking the answer.

Example:

```
       7,532  →   8                  54,321  →   6
       - 643  →  -4                 - 7,450  →  - 7
       6,889      4                  46,871       -1 + 9  →  8
6 + 8 + 8 + 9  →  4   OK       4 + 6 + 8 + 7 + 1  →  8        OK
```

Which problems would receive Chet Checker's OK?

5. 6,987 →	6. 3,729 →	7. 8,653 →	8. 567 →
- 764 →	- 934 →	- 267 →	- 265 →
6,223	2,795	8,286	302

Name _____

Chet: The Product Checker

Chet Checker claims to be the fastest "answer checker" for multiplication problems. Chet adds the digits of the factors repeatedly until a single digit remains for each number. He counts 9s as 0s. He then adds the digits in the product the same way. If the product of the digits for the factors equals the digit for the given product, the problem receives Chet Checker's OK.

Example:
```
   4,370  → 14 →  5           14,964 → 15 →  6
  x  74   → 11 → x2            x 421  →  7 → x7
  323,380            10 → 1    6,298,844         42 → 6
```

$3 + 2 + 3 + 3 + 8 + 0 →$ 19 → 1   OK $\qquad$ $6 + 2 + 9 + 8 + 8 + 4 + 4 →$ 32 → 5   NO

Which problems would receive Chet Checker's OK?

1. 2,567 →	2. 953 →	3. 8,932 →	4. 765 →
x 211 →	x 654 →	x 1,164 →	x 298 →
541,637	613,262	10,396,848	217,970

5. 9,876 →	6. 12,345 →	7. 24,680 →	8. 2,738 →
x 543 →	x 678 →	x 1,357 →	x 111 →
5,362,668	8,369,810	33,490,760	303,918

© Instructional Fair • TS Denison   IF2506 Math Magic: Slick Tricks with Numbers

Multiplication Madness

Name _____

Professor Mad Math has a strange way of multiplying numbers. Notice how Professor Mad Math multiplies 20 x 34 and 75 x 13:

~~20~~	x	~~34~~		75	x	13
~~10~~	x	~~68~~		37	x	26
5	x	136		~~18~~	x	~~52~~
~~2~~	x	~~272~~		9	x	104
1	x	544		~~4~~	x	~~208~~
		680 CORRECT!		~~2~~	x	~~416~~
				1	x	832
						975 CORRECT!

When asked to explain the method, Professor Mad Math said, "I simply divide the left number by 2 and multiply the right number by 2 until I reach 1. I like odd numbers so I cross out the rows that have even numbers on the left. I add the remaining numbers on the right." When told that 5 divided by 2 equals 2½ not 2, Professor Mad Math responded, "Fractions don't bother me. I just make certain that I add only the numbers on the right that have odd numbers on the left. Odd numbers are so exciting!"

Use Professor Mad Math's technique for the following problems.
Remember: 1. Divide the left factor by 2 (drop any remainder).
            2. Multiply the right factor by 2.
            3. Repeat step 1 and step 2 until you reach 1 on the left.
            4. Cross out the rows that have even factors on the left.
            5. Add the factors on the right that have not been crossed out.
            6. Is the sum the product of the first two numbers?

    29 x 38   44 x 50   65 x 12   78 x 61   80 x 90

Multiplication Magic

Consider the following multiplication problems. Do you see a pattern?

15 x 15 = 225	number of hundreds:	1 x 2 = 2
25 x 25 = 625	number of hundreds:	2 x 3 = 6
35 x 35 = 1,225	number of hundreds:	3 x 4 = 12

Fill in the blanks by following the pattern:

1. 45 x 45 = 2,025   number of hundreds: _____ x _____ = _____

2. 55 x 55 = _____

3. 65 x 65 = _____

4. 75 x 75 = _____

5. 85 x 85 = _____

6. 95 x 95 = _____

Does the pattern extend to three-digit numbers ending in 5?

    105 x 105 = 11,025   number of hundreds: 10 x 11 = 110

7. 115 x 115 = _____

8. 125 x 125 = _____

9. 135 x 135 = _____

10. 145 x 145 = _____

Using the pattern, what would 205 x 205 equal? _____
Amaze your friends by "knowing" 305 x 305, 405 x 405, etc.

Name _____

Win Some, Lose Some Multiplication

Once you know "the trick," you will be able to do multiplication problems like the following without a calculator or even without paper and pencil:

28 x 32   92 x 88   33 x 27   23 x 17   54 x 46   26 x 34

Consider the problems shown.

28 x 32 = 896   or   900 - 4   92 x 88 = 8,096   or   8,100 - 4

33 x 27 = 891   or   900 - 9   23 x 17 = 391   or   400 - 9

THE TRICK:

28 x 32 = (30 - 2) x (30 + 2) = 900 - 4 = 896

92 x 88 = (90 + 2) x (90 - 2) = 8,100 - 4 = 8,096

33 x 27 = (30 + 3) x (30 - 3) = 900 - 9 = 891

23 x 17 = (20 + 3) x (20 - 3) = 400 - 9 = 391

COMPLETE THESE:

1. 31 x 29 = (30 + ____)(30 - ____) = 900 - 1 = 899

2. 18 x 22 = (20 - ____) (20 + ____) = 400 - 4 = ____

3. 37 x 43 = (40 - ____)(40 + ____) = 1,600 - ____ = ____

4. 94 x 86 = (____ + ____)(____ - ____) = 8,100 - ____ = ____

5. 75 x 85 = (____ - ____)(____ + ____) = ____ - ____ = ____

6. 46 x 34 = (____ + ____)(____ - ____) = ____ - ____ = ____

7. 57 x 43 = (____ + ____)(____ - ____) = ____ - ____ = ____

© Instructional Fair • TS Denison   IF2506 Math Magic: Slick Tricks with Numbers

Name _____

Pascal's Triangle

Blaise Pascal (1623–1662) invented the triangle given below. It contains many patterns. One of its uses is to calculate the coefficients of a binomial expansion.

```
                    1
                 1     1
              1     2     1
           1     3     3     1
        1     4     6     4     1
     1     5    10    10     5     1
```

1. Complete __ __ __ __ __ __ __

   __ __ __ __ __ __ __ __

2. Describe the pattern _____.

The triangle gives the coefficients of the expansion of $(x + y)^n$ (n + 1 is the row of Pascal's triangle that is used).

Examples:
$(x + y)^0 = 1$,	1st (0 + 1) row	coefficient of x^0 is 1
$(x + y)^1 = 1x + 1y$,	2nd (1 + 1) row	coefficients are 1 and 1
$(x + y)^2 = 1x^2 + 2xy + 1y^2$	3rd (2 + 1) row	coefficients are 1, 2, 1
$(x + y)^3 = 1x^3 + 3x^2y + 3xy^2 + 1y^3$	4th (3 + 1) row	coefficients are 1, 3, 3, 1

Use Pascal's triangle to complete the following:

3. $(x + y)^4 =$ ____$x^4 +$ ____$x^3y +$ ____$x^2y^2 +$ ____$xy^3 +$ ____y^4

4. $(x + y)^5 =$ ____$x^5 +$ ____$x^4y +$ ____$x^3y^2 +$ ____$x^2y^3 +$ ____$xy^4 +$ ____y^5

The exponents decrease for the first variable, n through 0, and the exponents increase for the second variable, 0 through n. Each term has an exponent sum of n.

5. $(x + y)^6 =$ ____$x^- +$ ____$x^-y^1 +$ ____$x^-y^- +$ ____$x^3y^3 +$ ____$x^-y^- +$ ____$xy^- +$ ____y^-

Name _____

Integer Trees

Consider $5(3x + 4) = 15x + 20$ and the integer tree:

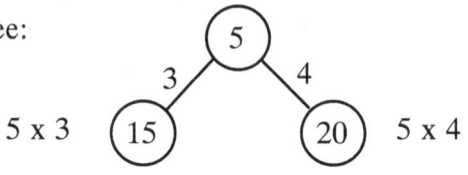

Consider $(2x + 5)(3x - 4) = 6x^2 + 7x - 20$ and the integer tree:

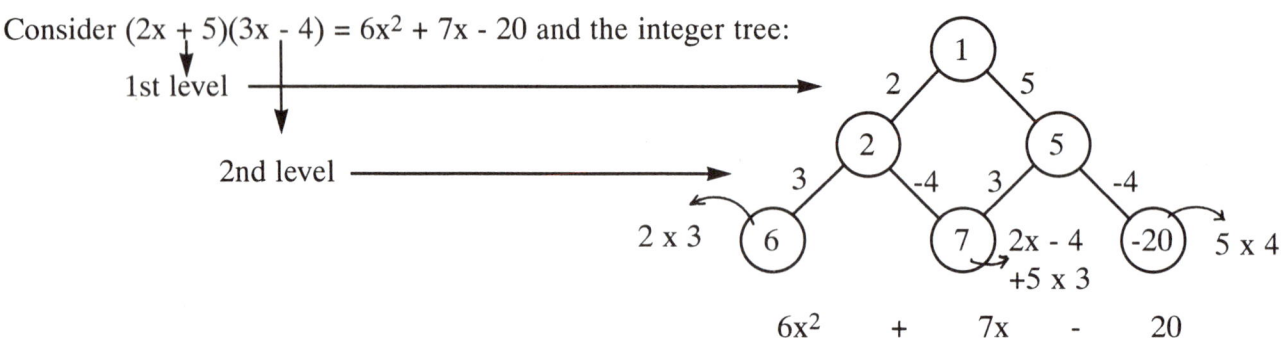

Integer trees can be used to find the product of several binomials:
$(4x - 1)(2x - 5)(3x + 7) = 24x^3 - 10x^2 - 139x + 35$

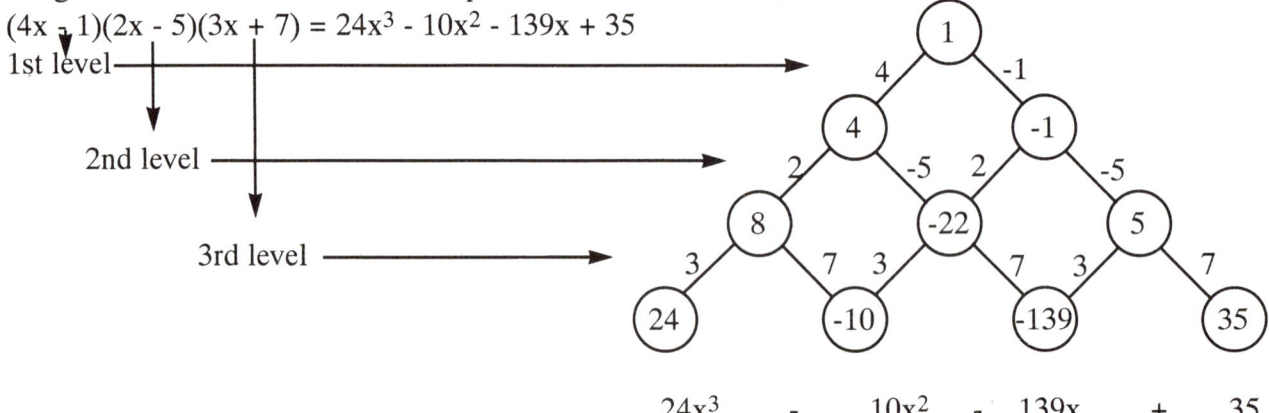

Complete the integer trees to compute the following products:

1. $(5x + 7)(3x - 1)$ 2. $(6x + 5)(x - 2)$ 3. $(2x + 3)(x - 1)(7x + 4)$

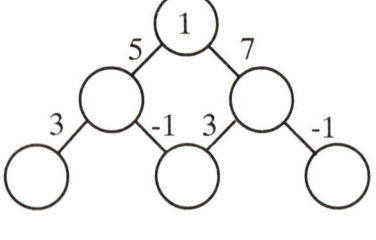 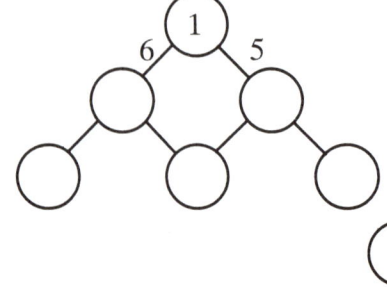 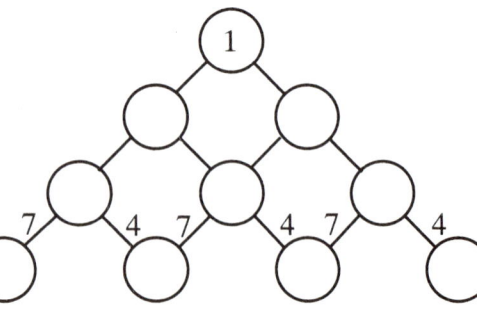

© Instructional Fair • TS Denison   IF2506 Math Magic: Slick Tricks with Numbers

Teacher Notes

The activities in "Magical Predictions" have predictable results based on algebraic identities, place value, factors, and the binary system. Some of the activities such as "Back to the Beginning" and "Fantastic Fractions" always result in the initial value chosen. Other activities such as "Double Vision," "Multi-vision," and "Caught in the Middle" result in numbers related to the initial choice. "High Noon" and "A Mouse in the House" have mathematical explanations for predicting the outcomes and are the type of activity used by magicians during "at-home audience" participation routines.

Back to the Beginning

Activity Students choose a number, perform various computations, and end with the original number.

Comments The title of the activity refers to the fact that the result of each problem is the number used at the beginning. These problems can be used as mental computation activities. If a scientific calculator is used, the equal sign should be used after each step so that the calculator's algebraic order of operations is not used.

To add some magic and amaze your students, do the activities orally and leave out the last step. When students give their results, perform the last step and "magically predict" their original numbers.

Have algebra students show how the activities work:

BTTB #1	BTTB #2	BTTB #3	BTTB #4
n	n	n	n
$2n$	$6n$	n^2	n^2
$2n+5$	$6n + 18$	$n^2 + 2n$	$n^2 + n$
$2(2n + 5) \to 4n + 10$	$(6n + 18)/3 \to 2n + 6$	$n^2 + 2n + 1$	$n^2 + n + 1$
$4n + 10 - 10 \to 4n$	$2n + 6 + 4 \to 2n + 10$	$(n^2 + 2n + 1) \to n + 1$	$(n^2 + n + 1)(n - 1) \to n^3 - 1$
$4n/2 \to 2n$	$(2n + 10)/2 \to n + 5$	$n + 1$	$n^3 - 1 + 1 \to n^3$
$2n/2 \to n$	$n + 5 - 5 \to n$	$n + 1 - 1 \to n$	$n^3/n \to n^2$
			$n^2/n \to n$

Age Predictor

Activity Students start with the month of their birth and end with a number representing their birth month and age.

Comments The result of this activity is the birth month followed by age. For example, 713 means the student's birthday is in July and the student is 13 years old. If a scientific calculator is used, the equal sign should be used after each step so that the algebraic order of operations is not used.

To amaze your students, leave out step 7. When students give their results, subtract 250 and "predict" the birth month and age.

m
2m
2m + 5
50(2m + 5) → 100m + 250
+ year of last birthday - year of birth = age(a). So 100m + 250 + a
(100m + 250 + a) - 250 → 100m + a
100m + a is the month of birth followed by age.

Birthday Predictor

Activity Students perform several computations and end with a number representing their birthday.

Comments The result of this activity is the number representation of the student's birthday. For example, 32,381 is a March 23, 1981 birthday.
For birthday m/d/y:

    m
    2m + 1
    50(2m + 1) → 100m + 50
    100m + 50 + d
    50(100m + 50 + d) → 5,000m + 2,500 + 50d
    2(5,000m + 2,500 + 50d) → 10,000m + 5,000 + 100d
    10,000m + 5,000 + 100d + y
    10,000m + 5,000 + 100d + y - 5,000 → 10,000m + 100d + y
    10,000m results in the month being the first two digits.
    100d results in the day being the middle two digits.
    y results in the year being the last two digits.

Pick a Number—Any Number

Activity Students choose a four-digit number and subtract reversals until a repeated digit occurs, always a 9 or 0.

Comments This makes an excellent calculator activity using the +/- key. If the difference in the display is smaller than the number written in reverse order, press the +/- key; then add the number written in reverse order. This emphasizes and reviews the fact that A - B = -B + A or A + (-B): to subtract, add the opposite. If the difference in the display is larger than the number written in reverse order, then simply subtract the smaller number.

Calculator Grid

Activity Students use a calculator keypad to choose a number. After following several steps, the student's result is always 198.

Comments For each case, n stands for the smallest number of the row, column, or diagonal.

Case 1: row
2. $100(n + 2) + 10(n + 1) + n = 111n + 210$
3. $111n + 210 - (100n + 10(n + 1) + (n+2)) =$
   $111n + 210 - (111n + 12) = 198$
4. $½(1 + 1) = 1$
5. $198 ÷ 1 = 198$

Case 2: column
2. $100(n + 6) + 10(n + 3) + n = 111n + 630$
3. $111n + 630 - (100n + 10(n + 3) + (n + 6)) =$
   $111n + 630 - (111n + 36) = 594$
4. $½(1 + 5) = 3$
5. $594 ÷ 3 = 198$

Case 3: diagonal
2. $100(n + 4) + 10(n + 2) + n = 111n + 420$
3. $111n + 420 - (100n + 10(n + 2) + (n + 4)) =$
   $111n + 420 - (111n + 24) = 396$
4. $½(1 + 3) = 2$
5. $396 ÷ 2 = 198$

NOTE: Step 3 is always a multiple of 198.

Double Vision

Activity Students write a three-digit number and after performing several multiplications have a six-digit product which is the first number repeated.

Comments Any three-digit number multiplied by 7, 11, and 13 results in the original number written twice as a six-digit number:
ABC x 7 x 11 x 13 = ABC x 1,001 = ABC,ABC

Example: 478 x 7 = 3,346; 3,346 x 11= 36,806; 36,806 x 13 = 478,478

A related activity involves writing any three-digit number twice to form a six-digit number. Then, divide the six-digit number by 7, 11, and 13 to obtain the original three-digit number: ABC,ABC ÷ 7 ÷ 11 ÷ 13 equals ABC, ABC ÷ (7 x 11 x 13) which equals ABC,ABC ÷ 1,001 or ABC, the original three-digit number.

Multi-vision

Activity Students choose a digit and after performing several multiplications have a six-digit product which is the chosen digit repeated six times.

Comments Any number 1 through 9, A, multiplied by 3 and 37,037 results in AAA,AAA: A x (3 x 37,037) = A x 111,111 = AAA,AAA.

Caught in the Middle

Activity Students add the digits of the given numbers repeatedly until a single digit remains. The result is the middle digit of the original number.

Comments Each of the numbers is written so that the sum of digits except for the middle digit equals a multiple of nine, 9N. Therefore, the sum of all the digits equals 9N + M where M is the middle digit.
9N + M = (10 - 1)N + M = 10N - 1N + M. When the digits of 10 are added to obtain 1, 10N - 1N + M becomes 1N - 1N + M = M, the middle digit.

Digit Predictor

Activity Students predict a digit chosen by other students.

Comments The difference of a number and the sum of its digits is a multiple of 9. For a four-digit number, $1{,}000t + 100H + 10T + U - (t + H + T + U) = 999t + 99H + 9T = 9(111t + 11H + T)$ which is a multiple of 9. The argument can be extended to any number.

Examples: 76 - 13 = 63 or 7 x 9   84 - 12 = 72 or 8 x 9

358 - 16 = 342 or 38 x 9   591 - 15 = 576 or 64 x 9

Once you know the remaining digits (after a digit is crossed out), add the digits. The sum of ALL the digits must be a multiple of 9. Subtract your sum from the next multiple of 9. The result is the crossed out digit. If the sum is a multiple of 9, the crossed out digit is 9.

Sum Predictor

Activity Students predict the sum of five four-digit numbers when given the first four-digit number.

Comments Subtracting 2 and writing a 2 in front of the difference is the same as adding 19,998 (20,000 - 2) to the first number.

Consider the following example:

First four-digit number:	7,491	Sum Prediction: 27,489
Second four-digit number:	5,893	The third number and the second number must equal 9,999.
Third four-digit number:	4,106	
Fourth four-digit number:	7,246	The fifth number and the fourth number must equal 9,999.
Fifth four-digit number:	2,753	
SUM	27,489	First number + 9,999 + 9,999 or first number + 19,998

Human Calculator

Activity Students choose five numbers from an array of 25 numbers, follow a set of directions, and predict the sum of the five numbers.

Comments Students choose one number from each of five columns. The numbers in each column are related. The middle digits are the same and the sum of the first and last digits are the same. In each column, the middle digits are 2, 5, 4, 3, and 6, respectively. Since only one number is chosen from each column, each of the middle digits appears in the tens place of the list of five numbers once. Their sum is 20 resulting in a 2 carried over to the hundreds place and 0 in the tens place. The sum of the last digits of the five numbers is then the same as the last two digits of the sum of the five chosen numbers.

In each column, the sum of the first and last digits is 8, 6, 10, 5, and 9, respectively. Their sum is 38 which added to the carried over 2 results in 40. Subtracting the sum of the last digits from 40 gives the first two digits of the final sum.

To make another human calculator array:
1. Choose five digits that have a sum which is a multiple of ten (10A).
2. Use these digits as the middle digits of each column.
3. Next choose five numbers that have a sum which when added to A give a multiple of ten. These five numbers will be the sum of the first and last digits of each number in the column.
4. After someone selects a number from each column, add the last digits and subtract from the multiple of ten from step 3. The predicted sum is that difference followed by the sum of the last digits.

Ex. Use 7, 8, 5, 4, and 6 for the middle digits. The sum is 30, so A = 3. Choose 10, 6, 8, 11, and 12 as the sums of the first and last digits. Their sum is 47 which added to A = 3 gives 50.

476	86	454	447	765
575	284	157	249	864
971	581	58	546	468
278	680	553	348	963
674	185	751	744	567

Choose: 575, 284, 454, 348, 963. Last digit sum = 24 and 50 - 24 = 26 so the sum = 2,624.

Fantastic Fractions

Activity Students choose two digits to form a fraction. Repeated addition and division steps result in the original numbers chosen.

Comments "Fantastic Fractions" provides practice in adding fractions, adding whole numbers and fractions, and dividing fractions. If the two numbers are X and Y, the steps result in the following:

Example: Results:
4   1. X

9   2. Y

$\frac{5}{2}$ 3. $(Y + 1) \div X = \dfrac{Y + 1}{X}$

$\frac{7}{18}$ 4. $\left(\dfrac{Y + 1}{X} + 1\right) \div Y = \dfrac{Y + 1 + X}{X} \times \dfrac{1}{Y} = \dfrac{Y + 1 + X}{XY}$

$\frac{5}{9}$ 5. $\left(\dfrac{Y + 1 + X}{XY} + 1\right) \div \dfrac{Y + 1}{X} = \dfrac{Y + 1 + X + XY}{XY} \times \dfrac{X}{(Y+1)} =$

$\dfrac{(Y + 1) + X(1 + Y)}{Y(Y+1)} = \dfrac{1 + X}{Y}$

4   6. $\left(\dfrac{1 + X}{Y} + 1\right) \div \dfrac{Y + 1 + X}{XY} = \dfrac{1 + X + Y}{Y} \times \dfrac{XY}{(Y + 1 + X)} = X$

9   7. $(X + 1) \div \dfrac{(1 + X)}{Y} = (X + 1) \times \dfrac{Y}{(X + 1)} = Y$

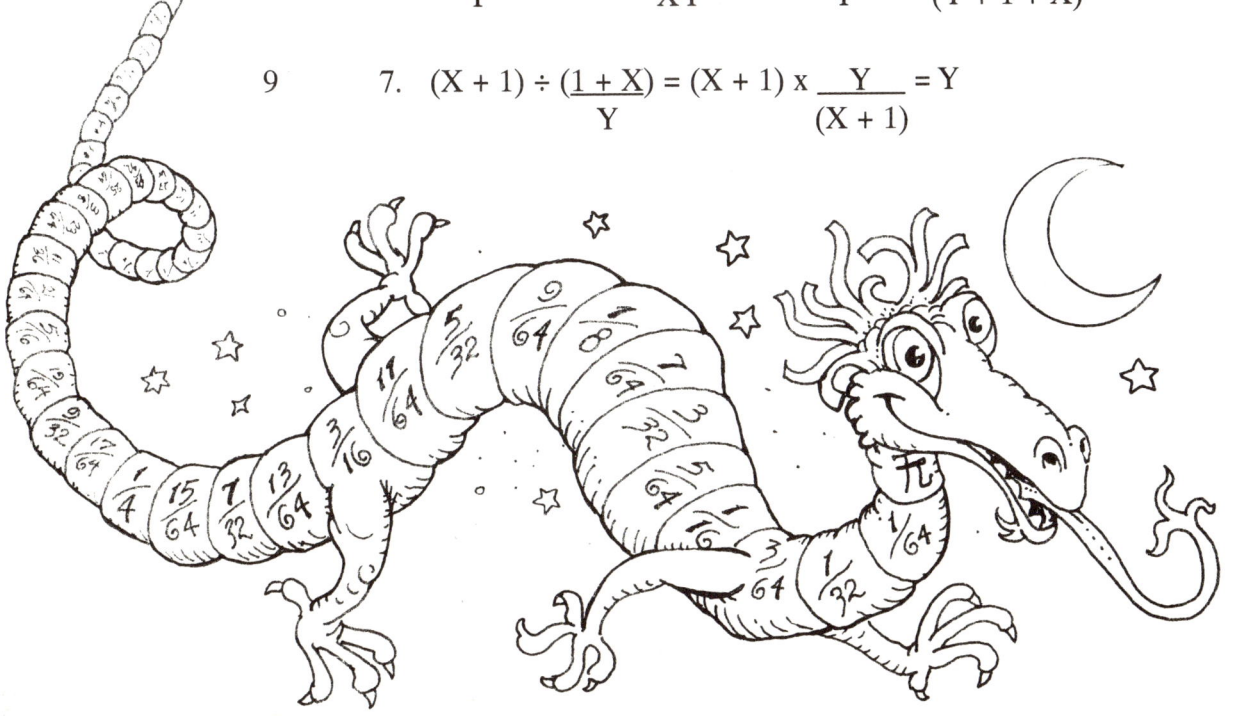

High Noon

Activity Students choose a number from a clock face, follow a series of directions, and ALWAYS end at 12:00—HIGH NOON!

Comments Once all numbers but 1, 4, 9, and 12 are crossed out in step 4, in order to end on **twelve,** the player must be on **one** (a count of 3—4, 8, 12) after Step 3. Any number on the clock as a first choice will result in landing on **one** at the end of three steps.

Consider the 12 possible cases.

Choice:	First #:	Second #:	Third #:	
One→	Three→	Eight→	One	3 + 5 + 5 = 13 (1)
Two→	Three→	Eight→	One	3 + 5 + 5 = 13 (1)
Three→	Five→	Nine→	One	5 + 4 + 4 = 13 (1)
Four→	Four→	Eight→	One	4 + 4 + 5 = 13 (1)
Five→	Four→	Eight→	One	4 + 4 + 5 = 13 (1)
Six→	Three→	Eight→	One	3 + 5 + 5 = 13 (1)
Seven→	Five→	Nine→	One	5 + 4 + 4 = 13 (1)
Eight→	Five→	Nine→	One	5 + 4 + 4 = 13 (1)
Nine→	Four→	Eight→	One	4 + 4 + 5 = 13 (1)
Ten→	Three→	Eight→	One	3 + 5 + 5 = 13 (1)
Eleven→	Six→	Nine→	One	6 + 3 + 4 = 13 (1)
Twelve→	Six→	Nine→	One	6 + 3 + 4 = 13 (1)

By changing step 4 and the directions in step 5, the final number can be changed to any number on the clock.

This type of activity is used by television magicians during at-home audience participation tricks.

A Mouse in the House

Activity Students track Max the Mouse through the floor plan of a house following a set of directions. Regardless of the initial starting point meeting certain conditions, Max ends up in the same room.

Comments The magic is that the student decides the starting point—any of five rooms in a nine-room house. No matter where the mouse is at the beginning, it ends up in the same room, the laundry room.

The trick can be explained by using a binary system of numbering the rooms or giving the rooms odd or even designations.
Consider the rooms to be numbered as shown:

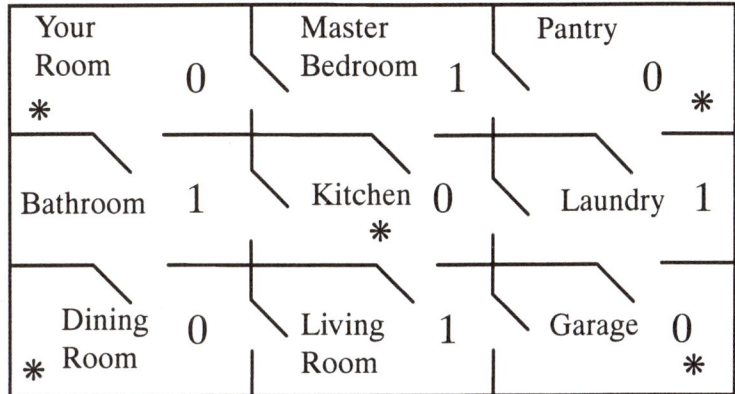

Starting in an *- marked room means Max starts in a room numbered 0. You hear one squeak. Max moves to a 1-room. You can close off your room because it is a 0-room. You hear two squeaks which means Max has moved "twice" and is again in a 1-room. You can close off the pantry and the dining room, both 0-rooms. This continues until Max is either in the kitchen or the garage. You hear one squeak which moves him into the laundry room. You close off the kitchen and garage and end the hunt for Max. No matter where Max starts, he will end up in the laundry room if the given directions are followed.

The order of the room closings can be changed as long as the kitchen is left to the end so that Max always has a way to move out of a room.

This is the type of activity that magicians will use to "read the minds" of at-home audiences.

Back to the Beginning

BTTB #1

1. Count the number of letters in your last name. _____
2. Double the number. _____
3. Add 5. _____
4. Double the result. _____
5. Take away 10. _____
6. Divide by 2. _____
7. Divide by 2 again. _____

Try the number of letters in your first name.

BTTB #2

1. Choose a number between 1 and 10. _____
2. Multiply by 6. _____
3. Add 18. _____
4. Divide by 3. _____
5. Add 4. _____
6. Divide by 2. _____
7. Subtract 5. _____

Try another number.

Name _____

BTTB #3

1. Choose a number between 1 and 10. _____
2. Square the number. _____
3. Add twice the original number. _____
4. Add 1. _____
5. Divide by 1 more than the original number. _____
6. Subtract 1. _____

Try another number. _____

BTTB #4

1. Choose a number between 1 and 10. _____
2. Square the number. _____
3. Add the original number. _____
4. Add 1. _____
5. Multiply by 1 less than the original number. _____
6. Add 1. _____
7. Divide by the original number. _____
8. Divide by the original number again. _____

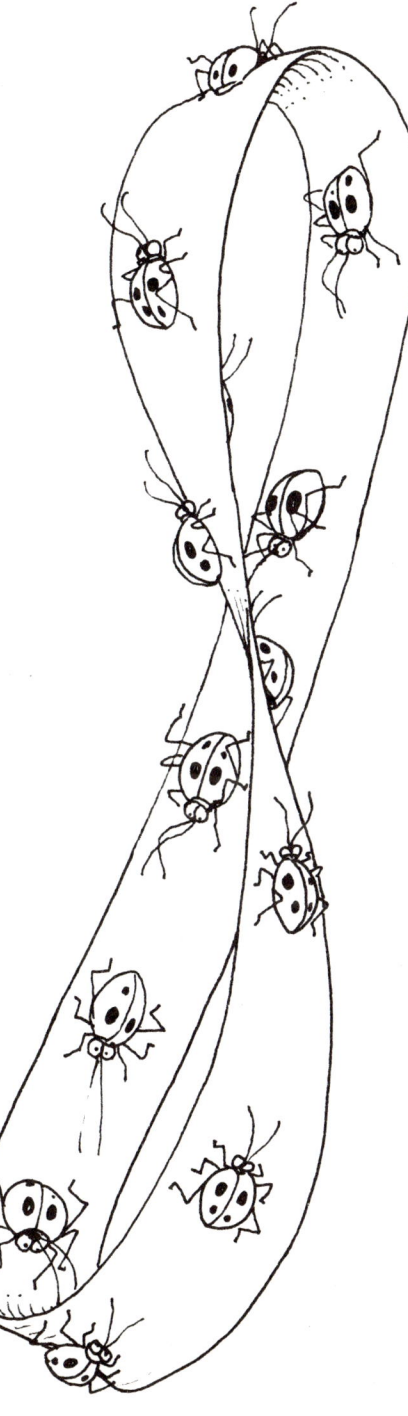

Name _____

Age Predictor

1. Start with the number of your birth month (ex. June = 6). _____

2. Double the number. _____

3. Add 5. _____

4. Multiply by 50. _____

5. Add the year of your last birthday. _____

6. Subtract the year of your birth. _____

7. Subtract 250. _____

8. Write your birth month number followed by your age. _____ Compare.

Name _____

Birthday Predictor

1. Start with the number of your birth month (ex. June = 6). _____

2. Double the number and add 1. _____

3. Multiply by 50. _____

4. Add the day of the month of your birthday. _____

5. Multiply by 50 again. _____

6. Multiply by 2. _____

7. Add the number formed by the last two digits of the year of your birth. _____

8. Subtract 5,000. _____

Name _____

Pick a Number—Any Number

1. Choose any four-digit number (any digit used only once). _____
2. Write the number in reverse order. _____
3. Subtract the smaller from the larger. _____
4. Repeat steps 1–3 until the digits in the result are the same. _____
5. Write the result. _____

Try another four-digit number. Write the result. _____

Name _____

Calculator Grid

1. Choose any row, column, or diagonal of your calculator containing numbers 1 through 9. 1. _____
2. Write the digits in reverse order. 2. _____
3. Subtract the smaller of step 1 and step 2 from the larger. 3. _____
4. Calculate half of one more than the hundreds digit. 4. _____
5. Divide step 3 by step 4. 5. _____

Repeat steps 1 through 5 using another set of digits. Compare the results.

1. ____ 2. ____ 3. ____ 4. ____ 5. ____

© Instructional Fair • TS Denison   41   IF2506 Math Magic: Slick Tricks with Numbers

Name _____

Double Vision

1. Choose any three-digit number. _____

2. Multiply by 7. _____

3. Multiply by 11. _____

4. Multiply by 13. _____

5. Are you seeing double? Explain the title. _____

   Try another three-digit number.   1. _____

   2. _____

   3. _____

   4. _____

Name _____

Multi-vision

1. Choose any number 1 through 9. _____

2. Multiply your choice by 3. _____

3. Multiply by 37,037. _____

   Try another number 1 through 9.   1. _____

   2. _____

   3. _____

4. Explain the title of the activity. _____

Name _____

Caught in the Middle

For each number below, add the digits together. If the sum has more than one digit, add the digits of the sum until a single digit remains.

1. 36,845 sum = _____
2. 27,154 sum = _____
3. 84,787 sum = _____

4. 3,586,524 sum = _____
5. 4,321,756 sum = _____
6. 2,527,153 sum = _____

7. 123,495,678 sum = _____
8. 346,745,182 sum = _____
9. 2,222,777 sum = _____

What do all the numbers have in common? _____
The title of the activity is a clue.

Name _____

Digit Predictor

Amaze a friend by predicting a number of your friend's choosing.

Have your friend do the following:
1. Write down any number that does not contain a zero.
2. Add the digits of the number and subtract that sum from the original number.
3. Cross out any digit of the difference.
4. Tell you the remaining digits.

You do the following:
5. Add the digits your friend tells you.
6. Subtract the sum from the next greater multiple of 9.
7. Have the difference—the "crossed out digit."

Your friend will be amazed! Try again with another number.

© Instructional Fair • TS Denison   43   IF2506 Math Magic: Slick Tricks with Numbers

Name _____

Sum Predictor

This trick lets you predict the sum of five four-digit numbers after only the first number is given.

Example:

1. Have someone write a four-digit number BETWEEN 0,000 and 9,999:

   3,476

2. Subtract 2 and place it in front of the result— 23,474.
   Write your prediction where no one can see it.

3. Have someone write another four-digit number under the first.

   5,648

4. Write a third number so that each digit added to the one above equals 9.

   4,351

5. Have someone write another four-digit number.

   7,012

6. Write a fifth number so that each digit added to the one above equals 9.

   2,987

7. Have someone find the sum of the five numbers.

   +_____
   23,474

8. Show your prediction—AMAZING!

Another example:

Someone	1.	7,491
You	2. Predict:	27,489
Someone	3.	5,893
You	4.	4,106
Someone	5.	7,246
You	6.	2,753
SUM	7.	27,489

Try the trick with other numbers.

Name _____

Human Calculator

1. Choose one number from each column below. _____, _____, _____, _____, _____
2. Add the last digits of the five numbers. _____
3. Subtract the sum from 40. _____
4. Write the difference followed by the sum. _____
5. Calculate the sum of the five numbers chosen. _____
6. Compare step 5 with step 4. _____
7. Repeat steps 1–6 with another set of five numbers.
8. Amaze your friends. Have them choose five numbers and find the sum. You quickly do steps 2-6 and predict their sum.

28	353	347	530	465
226	155	545	233	861
127	56	347	134	366
325	254	248	35	267
424	452	446	332	69

FIRST TRY:
Step 1:   Step 2:   Step 3:
(numbers)   (last digits)
                                40 -   =

                                Step 4:

+ _____ + _____   ___ ___ ___

SECOND TRY:
Step 1:   Step 2:   Step 3:
(numbers)   (last digits)
                                40 -   =

                                Step 4:

+ _____ + _____   ___ ___ ___

© Instructional Fair • TS Denison   IF2506 Math Magic: Slick Tricks with Numbers

Fantastic Fractions

Name _____

The following activity not only provides practice in addition and division of fractions but has some *fantastic* results. Use the example to follow the steps. Try the activity using your own numbers.

Example:	Steps:	Your Results:
1. 4	1. Choose a digit 1 through 9.	1.
2. 9	2. Choose another digit 1 through 9.	2.
3. $(9 + 1) \div 4 =$ $5/2$	3. Add 1 to your second choice and divide by the first.	3.
4. $(5/2 + 1) \div 9 =$ $7/2 \times 1/9 =$ $7/18$	4. Add 1 to your third result and divide by the second.	4.
5. $(7/18 + 1) \div 5/2 =$ $25/18 \times 2/5 =$ $5/9$	5. Add 1 to your fourth result and divide by the third.	5.
6. $(5/9 + 1) \div 7/18 =$ $14/9 \times 18/7 =$ 4	6. Add 1 to your fifth result and divide by the fourth.	6.
7. $(4 + 1) \div 5/9 =$ $5 \times 9/5 =$ 9	7. Add 1 to your sixth result and divide by the fifth.	7.
8. THE SAME!!	8. Compare the sixth and seventh steps to the first and second steps. Surprised?	8.

Try the steps for other pairs of digits. See if you have the same results.

Name _____

HighNoon

Start at High Noon—12:00 and follow the steps below. Always move clockwise.

1. Choose any number on the clock and spell it out moving one hour for each letter. Example: Choose the number FIVE. You would spell it with F on the 1, I on the 2, V on the 3, and E on the 4.

2. Start with the number you landed on in step 1 and spell out that number so that you land on a second number.

3. Start with the number you landed on in step 2 and spell out that number so that you land on a third number.

4. Cross out 2, 3, 5, 6, 7, 8, 10, and 11.

5. Spell the number you landed on in step 3, causing you to land on a fourth and final number—High Noon!

Start at 12:00 and repeat steps 1–5, choosing a different starting number for step 1. What is your fourth and final number? _____

Name _____

A Mouse in the House

Max the Mouse has escaped from his cage in the laundry room. You know he will be in any room that might have food in it: kitchen, pantry, dining room, garage where his food is stored and, of course, your bedroom. When he runs through doorways, you hear a squeaky noise. Assume Max is in any room of the house that might have food. Use the floor plan below. Follow the directions to catch Max.

Remember: You hear a squeak as he passes through a doorway.
He can only travel through open doors. Once you have closed the doors to a room, he cannot enter it.
Choose **any** room that might have food in it (*) as a starting point.

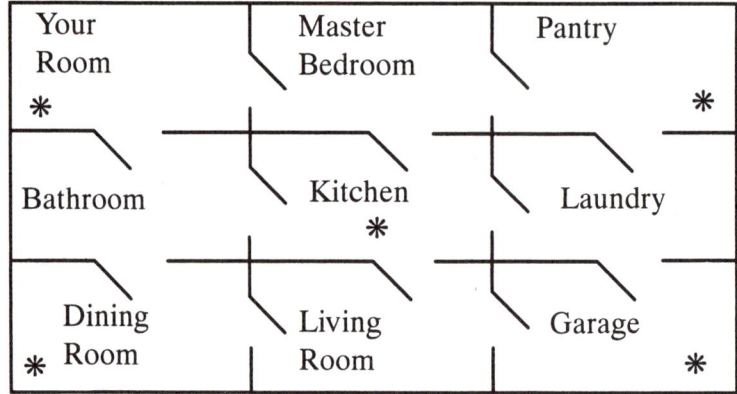

1. You hear one squeak. Max has passed through one doorway to another room.
2. After checking your room, you close both doors to your room.
3. You hear two squeaks. Max has passed through two doorways.
4. You check the pantry and the dining room, and close all doors to those rooms.
5. You hear one squeak. Max has passed through one doorway.
6. You close the doors to the master bedroom.
7. You hear two squeaks. Max has again passed through two doorways.
8. You close off the living room and the bathroom.
9. You hear one last squeak.
10. You think you know where he is. You close off the kitchen and the garage.
11. You find Max in the laundry room sitting next to his cage waiting for his lunch.
12. Have Max start in another room. Repeat the steps and see where he ends up.

Magic Squares, Triangles, and Circles

Teacher Notes

A **magic square** is a square array of numbers in which the sum of the numbers in each column, row, or diagonal, the magic sum, is the same. A **magic triangle** has numbers along the three sides and at the vertices of a triangle so the sums of the numbers along each side are equal. A **magic circle** is a circular arrangement of numbers in which the sums of the numbers along the diameters are equal.

Magic squares can be used as self-checking assignments. Each cell of the magic square contains a problem. The answers to the problems form a magic square. Students can check their answers by finding the sums of the rows, columns, and diagonals. If the sums are not equal, the intersection of the row and column having different sums is an incorrect answer.

Magic Squares
Odd Magic Squares

An odd magic square—3 x 3, 5 x 5, etc.—can be made by following the steps below.

1. The first number is placed in the top row, middle square.

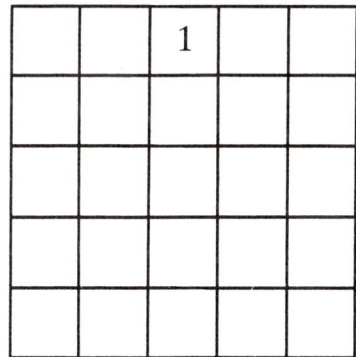

2. Always try to place the next number diagonally up and to the right of the last number.

Steps 3–6 tell you what to do when step 2 cannot be followed.

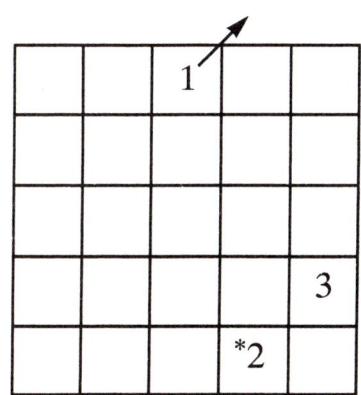

3. If trying to do step 2 results in being above the magic square, place the number in the bottom square of the next column to the right.*

49

4. If trying to do step 2 results in being to the right of the magic square, place the number in the left hand square of the row above the last placed number. *

		1		
*4				
				3 ↗
			2	

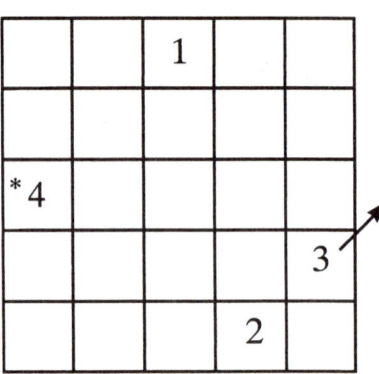

5. If the square diagonally up and to the right is already filled, place the number in the square immediately below the last placed number.*

		1	8	15
	5 ↗	7	14	
4	*6	13		
10	12			3
11			2	9

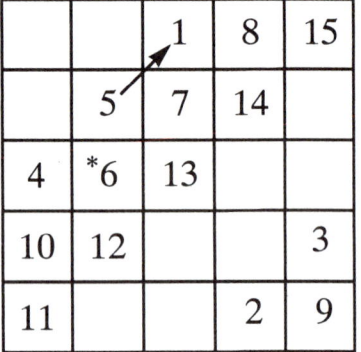

6. After filling the upper right-hand square, place the next number immediately below the upper right-hand square.*

17	24	1	8	15 ↗
23	5	7	14	*16
4	6	13	20	22
10	12	19	21	3
11	18	25	2	9

The example given uses the numbers 1, 2, 3, . . . 24, 25. Any series of numbers can be used as long as the difference between terms is constant. Some series are

0, 3, 6, 9, 12, 15, . . . 10, 12, 14, 16, 18, . . .
-5, -3, -1, 1, 3, 5, 7, . . . -25, -22, -19, -16, -13, . . .

To use fractions or decimals, divide the numbers of a sequence by a constant.

Example: Divide the series 1, 2, 3, . . . 25 by 8.

$2\frac{1}{8}$	3	$\frac{1}{8}$	1	$1\frac{7}{8}$
$2\frac{7}{8}$	$\frac{5}{8}$	$\frac{7}{8}$	$1\frac{3}{4}$	2
$\frac{1}{2}$	$\frac{3}{4}$	$1\frac{5}{8}$	$2\frac{1}{2}$	$2\frac{3}{4}$
$1\frac{1}{4}$	$1\frac{1}{2}$	$2\frac{3}{8}$	$2\frac{5}{3}$	$\frac{3}{8}$
$1\frac{3}{8}$	$2\frac{1}{4}$	$3\frac{1}{8}$	$\frac{1}{4}$	$1\frac{1}{8}$

Even Magic Squares

A 4 x 4 magic square with a given sum can be made from the following model:

The magic square to the right results in a magic sum of K + 20.

K	1	12	7
11	8	K−1	2
5	10	3	K+2
4	K+1	6	9

Example: Sum of 34 (K = 14 from 34 − 20)

14	1	12	7
11	8	13	2
5	10	3	16
4	15	6	9

To have fraction or decimal numbers, divide each square by a constant.

Example: Each number in the square above was divided by 25 to obtain this square.

0.56	0.04	0.48	0.28
0.44	0.32	0.52	0.08
0.2	0.4	0.12	0.64
0.16	0.6	0.24	0.36

Sum: 1.36

Magic Squares

Activity Students use the digits 1 through 9 to create a 3 x 3 magic square and find the magic sum. Students then complete a 4 x 4 magic square by first calculating the magic sum from one completed diagonal.

Comments The magic sum provides the information needed for students to complete the square.

Fraction Magic Squares

Activity Students solve fraction problems to create a 4 x 4 magic square and calculate the magic sum. Students then complete a 4 x 4 magic square of fractional numbers by first calculating the magic sum from one completed diagonal.

Comments All four operations are included. The first magic square uses ⅛s. The second magic square uses ⅒s. The magic sum provides the information needed for students to complete the square.

Decimal Magic Squares

Activity Students solve decimal problems to create a 4 x 4 magic square and calculate the magic sum. Students then complete a 4 x 4 magic square of decimal numbers by first calculating the magic sum from one completed diagonal.

Comments All four operations are included. The magic sum provides the information needed for students to complete the square.

Equation Magic Square

Activity Students solve multi-step equations to complete a 5 x 5 magic square and calculate the magic sum. Students should calculate the sums of all rows, columns, and diagonals as a check of their solutions.

Comments All four operations are included. Some equations require collecting variable terms to one side.

A Magical Magic Square

Activity Students complete a 4 x 4 magic square, calculate the magic sum, and find the sums of particular parts of the magic square. The sums provide some interesting results.

Comments The magic square was formed by counting backwards and filling in the squares on the two diagonals.

16	15	14	13
12	11	10	9
8	7	6	5
4	3	2	1

The remaining squares were filled in by counting forward.

16	2	3	13
5	11	10	8
9	7	6	12
4	14	15	1

This magic square is similar to Durer's 1514 engraving *Melancholia* except the two middle columns are interchanged. The *Melancholia*, to the right, shows the date (1514) of the engraving in the bottom line.

16	3	2	13
5	10	11	8
9	6	7	12
4	15	14	1

Magic Square Triangle

Activity Students complete three magic squares, a 3 x 3, a 4 x 4, and a 5 x 5, arranged along the sides of a triangle and calculate the magic sums of 9, 16, and 25 respectively. Students are asked to respond to a question concerning 3-4-5 triangles.

Comments A 3-4-5 triangle is a right triangle since $3^2 + 4^2 = 5^2$ (9 + 16 = 25). The Pythagorean Theorem states that the sum of the squares of the legs of a right triangle equals the square of the hypotenuse. Any three numbers a, b, and c that satisfy $a^2 + b^2 = c^2$ is called a Pythagorean triple. Another commonly used triple is 5-12-13.

Any multiple of a Pythagorean triple, such as 6-8-10, 9-12-15, and 12-16-20 which are multiples of the 3-4-5 triple, is also a Pythagorean triple. The numbers can be the measures of the three sides of a right triangle.

Proof: Given three numbers a, b, and c, and a constant K, K ≠ 0.
If a-b-c is a Pythagorean triple, then Ka-Kb-Kc is a Pythagorean triple.

$a^2 + b^2 = c^2$ given
$(Ka)^2 + (Kb)^2 = K^2a^2 + K^2b^2 =$ rules of exponents
$\qquad K^2(a^2 + b^2) =$ factoring
$\qquad K^2c^2$ substitution $a^2 + b^2 = c^2$
$\qquad (Kc)^2$ rules of exponents
Therefore, $(Ka)^2 + (Kb)^2 = (Kc)^2$
Ka-Kb-Kc is a Pythagorean triple.

Magic Triangles

Activity Students complete magic triangles to obtain a given magic sum. Students are asked to create additional magic triangles and to determine the largest magic sum for digits 1 through 9.

Comments The sum of all three sides of any magic triangle equals three times the magic sum. To obtain the smallest (largest) magic sum, place the smallest (largest) numbers at the vertices.

For digits 1–6:
The smallest magic sum is $[2(1) + 2(2) + 2(3) + 4 + 5 + 6] \div 3 = 27 \div 3 = 9$.
The largest magic sum is $[2(4) + 2(5) + 2(6) + 1 + 2 + 3] \div 3 = 36 \div 3 = 12$.
The only possible magic sums for digits 1–6 are 9, 10, 11, and 12.

For digits 1–9:
The largest magic sum is
$[2(7) + 2(8) + 2(9) + 1 + 2 + 3 + 4 + 5 + 6] \div 3 = 69 \div 3 = 23$.

Magic Circles

Activity Students complete four magic circles using given number sets and magic sums. Students are asked to determine a relationship between the magic sum and the numbers used.

Comments When the numbers used to complete a magic circle are consecutive with a set interval between the numbers, the magic sum is 3 times the middle term. Some students may conclude that the magic sum is 1½ times the sum of the smallest and largest numbers. This may be easier to compute since it does not require finding the median. One and a half times the sum of the smallest, S, and largest, L, numbers is equal to 3 times the average of the smallest and largest number (or 3 times the median).

$$(1\tfrac{1}{2})(S + L) = (\tfrac{3}{2})(S + L) = 3 \underline{(S + L)} \text{ equivalent to 3 times the median}$$
$$\phantom{(1\tfrac{1}{2})(S + L) = (\tfrac{3}{2})(S + L) = 3\ } 2$$

A follow-up activity would be calculating the magic sum given a certain series of numbers:

    Consecutive integers 1–21. Magic Sum = ?   (33)
    First 13 positive multiples of 6. Magic Sum = ?   (126)

Name _____

Magic Squares

A **magic square** is a square arrangement of numbers in which the sum of the numbers in each column, row, or diagonal, the magic sum, is the same.

Arrange the digits 1 through 9 (using each digit only once) so that the sum of each column, row, and diagonal is the same. What is the magic sum? _____

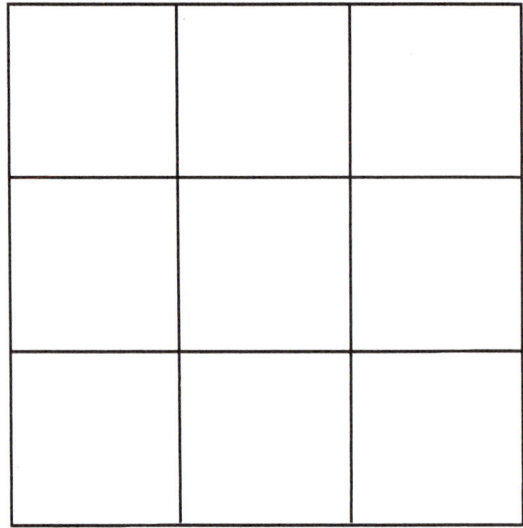

Complete the following magic square. What is the magic sum? _____

	2		14
	16	26	
10	20		32
8			18

© Instructional Fair • TS Denison   IF2506 Math Magic: Slick Tricks with Numbers

Fraction Magic Squares

Name _____

Solve the following computation problems. The answers should result in a magic square. What is the magic sum? _____

⅞ + ⅞	½ × ¼	2 × ¾	½ + ⅜
⅝ + ¾	¾ ÷ ¾	¾ + ⅞	½ × ½
⅞ − ¼	¾ + ½	¾ ÷ 2	⅔ ÷ ⅓
¾ − ¼	2 − ⅛	⅜ + ⅜	⅞ + ¼

Complete the following magic square. What is the magic sum? _____

	⅒		⁷⁄₁₀
	⅘	1³⁄₁₀	
½	1		1⅗
⅖			⁹⁄₁₀

Decimal Magic Squares

Name _____

Solve the following computation problems. The answers should result in a magic square. What is the magic sum? _____

3 x 0.3	0.2 + 0.05	1 - 0.2	0.4 + 0.15
3 ÷ 4	0.8 - 0.2	1 - 0.15	0.9 ÷ 3
0.5 - 0.05	0.3 + 0.4	0.7 ÷ 2	0.7 ÷ 0.7
0.9 - 0.5	1 - 0.05	0.55 ÷ 1.1	1.3 x 0.5

Complete the following magic square. What is the magic sum? _____

0.7	0.05		
	0.4	0.65	0.1
		0.15	
0.2		0.3	0.45

Equation Magic Square

Name _____

Solve the following equations. The answers should result in a magic square. What is the magic sum? _____

½x + 3 = 20	2x + 4 = 100	6x + 1 = 4x + 5	8x = 6x + 32	⅓x - 1 = 9
x =	x =	x =	x =	x =
½x + 2 = 25	6x - 10 = 50	4x + 4 = 60	2x - 6 = 50	3x + 4 = 100
x =	x =	x =	x =	x =
3x - 4 = 20	5x = 24 + 3x	2x - 2 = 50	5x - 10 = 3x + 70	¼x - 1 = 10
x =	x =	x =	x =	x =
60 = 2x + 20	2x + 2 = 50	½x + 1 = 20	⅓x - 4 = 10	4x = x + 18
x =	x =	x =	x =	x =
30 = 3x - 36	½x - 3 = 15	⅖x = 10	5x - 5 = 15	2x = 36
x =	x =	x =	x =	x =

A Magical Magic Square

Name _____

Complete the following magic square and answer the questions below to discover some "magical" results.

16 A	2 B	C	13 D
5 E	F	10 G	8 H
I	7 J	K	L
4 M	N	O	1 P

1. What is the magic sum? _____

2. What is the sum of the four corner squares (A, D, M, and P)? _____

3. What is the sum of the four interior squares (F, G, J, and K)? _____

4. What is the sum of the top row middle and bottom row middle squares (B, C, N, and O)? _____

5. What is the sum of the middle squares of the side columns (E, I, H, and L)? _____

6. Compute the sum of the squares of the top two rows ($A^2 + B^2 + C^2 + \ldots + H^2$). _____

7. Compute the sum of the squares of the bottom two rows ($I^2 + J^2 + K^2 + \ldots + P^2$). _____

Name _____

Magic Square Triangle

Complete the magic square on each of the three sides of the triangle. Find the three magic sums. Do you know what makes a 3-4-5 triangle special?

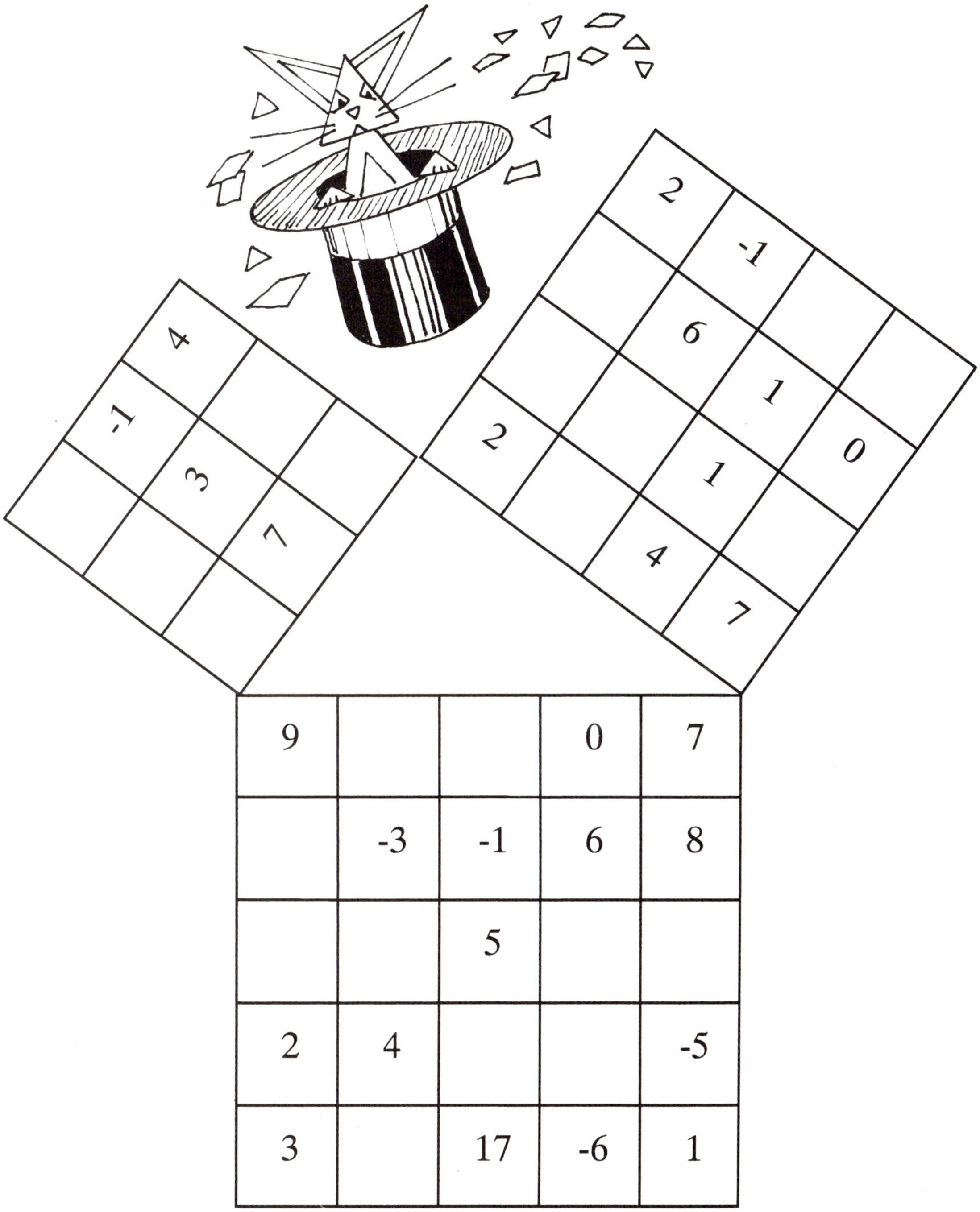

Name _____

Magic Triangles

A **magic triangle** is a triangular arrangement of numbers in which the sum of the numbers along each side, the magic sum, is the same.

Arrange the digits 1 through 6 in the six boxes of each triangle so the magic sum is:

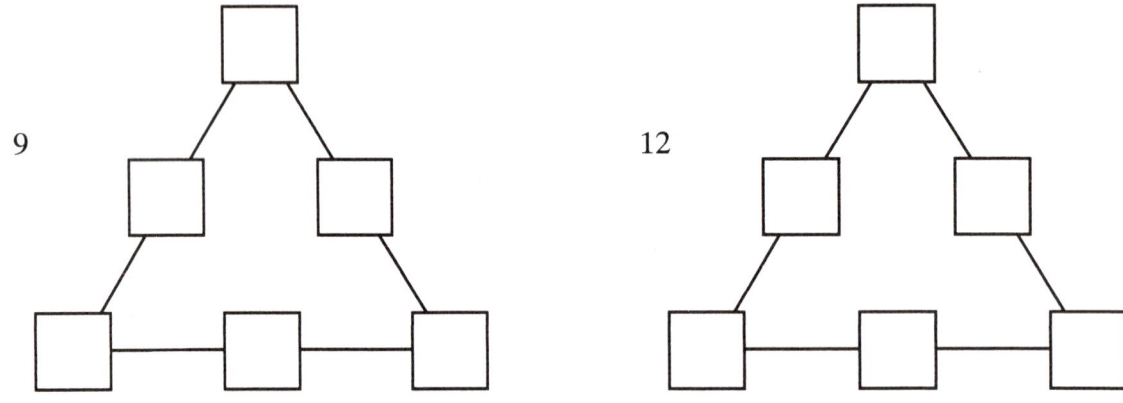

Can you find magic triangles, using 1 through 6, that have other magic sums? _____

Arrange the digits 1 through 9 (using each digit only once) so that the magic sum is:

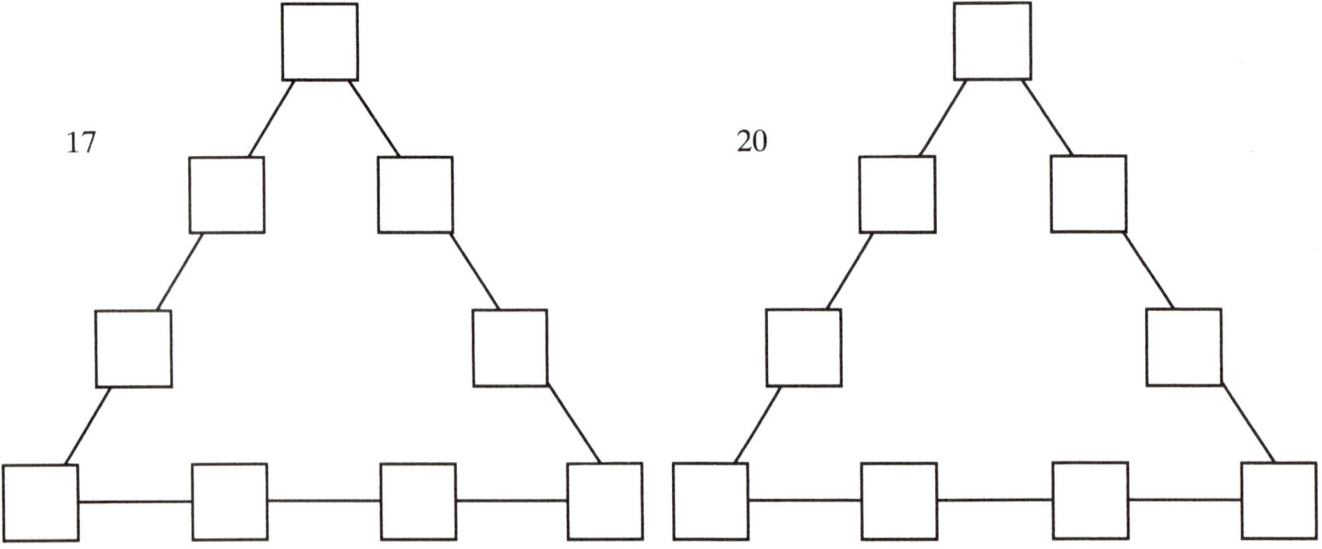

What is the largest magic sum of a magic triangle using digits 1 through 9? _____

© Instructional Fair • TS Denison 62 IF2506 Math Magic: Slick Tricks with Numbers

Magic Circles

Magic circles have the same sum along each line of circles. That sum is called the magic sum.

Use the digits 1 through 7.
Magic Sum = 12

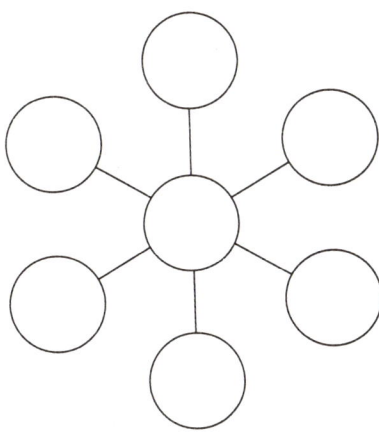

Use the odds 1 through 13.
Magic Sum = 21

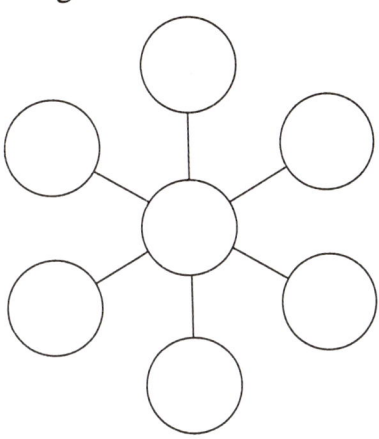

Use the first 9 positive multiples of 3.
Magic Sum = 45

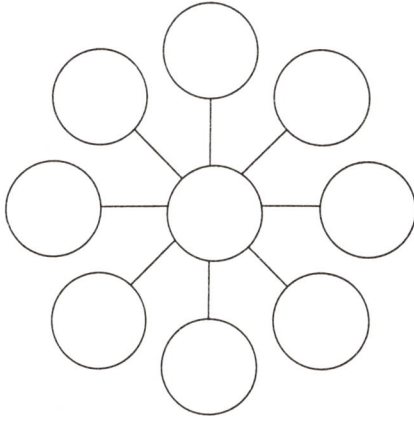

Use the first 11 positive multiples of 5.
Magic Sum = 90

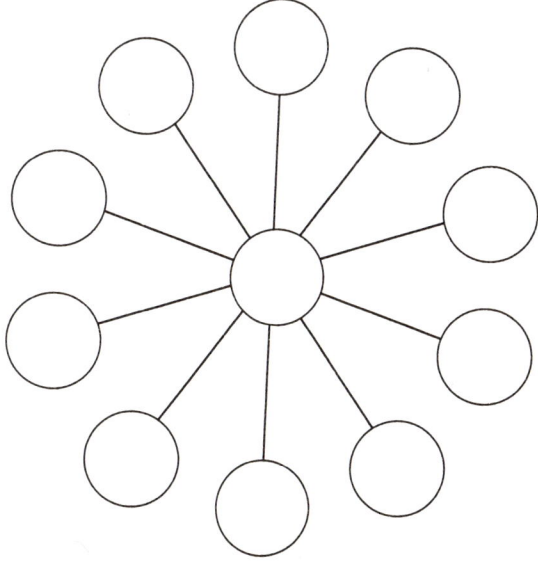

Can you find a rule that results in the magic sum given the numbers used?

Math Magic with Cards and Cups

Teacher Notes

The activities in this section use standard playing cards, student-made mind-reading cards, and cups. The magic tricks can be performed by the teacher as a class-wide activity or by students in small groups. The magic tricks are based on number theory concepts including place value, multiples, factors, even/odd, and binary numbers. The Comments provide explanations of the mathematics of the magic tricks. The mystery is explained while the magic of math continues as students perform the tricks.

Magic Card

Activity — A "magic card" is placed in a deck in the eighteenth position and then it appears at the top.

Materials — Standard deck of playing cards
Optional: a die and a 3½ ♥ card (see below)

Comments — When the deck is "cut," usually about 23 to 29 cards will remain in the top part. (If fewer than 20 cards remain, start over.) After the cards are counted out, the "magic card" is the eighteenth card from the BOTTOM. The sum of the digits of the number of cards remaining subtracted from the number of cards will equal 18 (as long as fewer than 30 cards remain). If 30 or more cards remain, then repeat, removing cards from the top.

When the sum of the digits of a two-digit number is subtracted from the number, the difference is a multiple of 9: $10T + U - (T + U) = 9T$.
For numbers 20 through 29, the difference is 18.

To add a twist to the activity:
Make a 3½ ♥ by writing a ½ and drawing half a heart on a 3 ♥. Use it as the "magic card." Before you reveal the "magic card," roll a die and predict that the card is the average of the numbers on the top and bottom of the die. The sum of the opposing numbers on a standard die is 7, so the average is always 3½. Students will think you have made a mistake until you reveal the 3½ ♥.

© Instructional Fair • TS Denison   IF2506 Math Magic: Slick Tricks with Numbers

Pick a Card—Any Card

Activity Students find a chosen card out of 27 after several deals.

Materials 27 cards from a standard deck of cards

Comments After the dealer places the "chosen card" column in the middle of the deck, the chosen card is in the middle third of the deck. The card is located in positions 10 through 18, after the second deal. The third time the cards are dealt, the chosen card is in positions 13 through 15. After the last pickup, the chosen card is in position 14. The given steps also work with 21 cards (3 columns of 7) with a final position of 11. If 33 cards (3 columns of 11) are used, do an additional deal for a final position of 17.

Magic Deal

Activity Students repeatedly deal cards until one is left, the chosen card.

Materials A deck of cards separated into a stack of 15 and a stack of the remaining cards

Comments The chosen card becomes the sixteenth card from the top of the smaller deck. For purposes of explanation, think of the cards being numbered 1, 2, 3, . . . , 14, 15, and 16—the chosen card added to the bottom of the deck. The MAGIC DEAL removes cards 1, 3, 5, 7, 9, 11, 13, and 15 the first time. Cards 2, 4, 6, 8, 10, 12, 14, and 16 remain in the deck. Then cards 2, 6, 10, and 14 are removed and cards 4, 8, 12, and 16 remain. Cards 4 and 12 are removed with 8 and 16 still in the deck. The last card to be removed is 8, so card 16, the chosen card, remains.

Card Flip

Activity Students predict whether a covered card is face up or face down.

Materials Three aces

Comments Every time two of the three cards are turned over, the number of aces may change, but the EVENNESS or ODDNESS of the number of aces stays the same.

Examples: ■ A A (EVEN) could become:
■ ■ ■, A■A, ■AA, or AA■ —ALL EVEN
■ A ■ (ODD) could become:
A■ ■, ■A ■, ■ ■ A, or A A A —ALL ODD

If the two cards revealed result in the EVENNESS or ODDNESS of aces at the beginning, the covered card must be face down (value of 0).
If the two cards revealed result in the opposite of what occurred at the beginning, the covered card must be face up (value of 1) to change EVEN to ODD, or ODD to EVEN.

Card Predictor

Activity Student uses two cards to predict the result of a number game.

Materials Deck of cards

Comments Let the card kept by the volunteer be V. Let the card you have seen be Y. Step 7 results in $5(2V + 2)$ or $10V + 10$. Step 8 is the difference $10 - Y$. Step 9 results in $10V + 10 - (10 - Y)$ which equals $10V + Y$. Therefore, the turned over cards will be the "answer" from step 9. Keep in mind if Y is a 10 or a face card, then it has a value of 0.

Three Card Guess

Activity Students predict the numbers of three cards after a series of math steps are performed.

Materials Three cards, ace through 9

Comments If the numbers on the cards are A, B, and C, the steps result in:
1. 2A
2. (2A + 5) x 5 = 10A + 25
3. 10A + 25 + B
4. (10A + 25 + B) x 10 = 100A + 250 + 10B
5. 100A + 250 + 10B + C

So 100A + 250 + 10B + C - 250 = 100A + 10B + C, the three-digit result gives the numbers A, B, and C on the cards.

Mind Reading Cards

Activity Students make a set of mind-reading cards. The cards are used by a student to *read the mind* of another student to predict a chosen number.

Materials 5 blank cards (large index cards) to make mind-reading cards

Comments The numbers in the upper left-hand corners of the mind-reading cards are the place values for *base 2:* 1, 2, 4, 8, and 16. Each number from 1 through 31 is written as the sum of 1, 2, 4, 8, and 16. The number is placed on the corresponding cards.

1 = 1	9 = 1 + 8	17 = 1 + 16	25 = 1 + 8 + 16
2 = 2	10 = 2 + 8	18 = 2 + 16	26 = 2 + 8 + 16
3 = 1 + 2	11 = 1 + 2 + 8	19 = 1 + 2 + 16	27 = 1 + 2 + 8 + 16
4 = 4	12 = 4 + 8	20 = 4 + 16	28 = 4 + 8 + 16
5 = 1 + 4	13 = 1 + 4 + 8	21 = 1 + 4 + 16	29 = 1 + 4 + 8 + 16
6 = 2 + 4	14 = 2 + 4 + 8	22 = 2 + 4 + 16	30 = 2 + 4 + 8 + 16
7 = 1 + 2 + 4	15 = 1 + 2 + 4 + 8	23 = 1 + 2 + 4 + 16	31 = 1 + 2 + 4 + 8 + 16
8 = 8	16 = 16	24 = 8 + 16	

Four Cups

Activity Students predict the location of an object after a series of hidden moves.

Materials 4 plastic cups
Solid object like a die

1   2   3   4

Comments Consider the cups to be numbered:

Mentally note whether the object starts in an EVEN or an ODD cup. Each time you hear the object drop into a cup, mentally change the even to odd or the odd to even (since the object goes into a cup NEXT to it). When the pouring is finished, if ODD, have the person remove cup 4 since the object is in cup 1 or cup 3. Have the person pour one more time so that the object is in cup 2—your prediction. When the pouring is finished, if EVEN, have cup 1 removed since the object is in cup 2 or cup 4. Have the person pour one more time so that the object is in cup 3—your prediction.

Three Cups

Activity Students predict the location of a hidden object after a series of hidden moves.

Materials 3 plastic cups of various colors or patterns

Comments Take note of the location (1, 2, or 3) of one of the cups before you turn around by color or label, etc. As the person tells you the moves (1 and 3, 2 and 3, etc.), KEEP TRACK of your cup's location. When you turn around, if your cup is not where it should be, then it was one of the empty ones moved at the start and the object is where your cup should be. If your cup is where it should be, then it contains the object.

Magic Card

1. Choose a card to be your magic card. Secretly place the card so that it is the eighteenth card from the top of a deck (deck face down).
2. Have someone "cut the deck in half."
3. Discard the bottom half.
4. Have someone count out the number of cards remaining. Be certain to physically count out the cards so that the top card face down is placed down first so it becomes the bottom card.
5. Have someone find the sum of the digits of the number cards counted.
6. Have someone remove that number of cards from the top of the deck.
7. Make up a story about your magic card.
8. Turn over the top card to reveal the card.

Pick a Card—Any Card

1. Use 27 cards. Fan them out face down and have someone choose a card, look at it (NOT showing you), return the card to the deck, and shuffle the deck.
2. Take the cards and lay them out one at a time face up into three columns:

1	2	3
4	5	6
7	8	9
10	11	12
13	14	15
16	17	18
19	20	21
22	23	24
25	26	27

The diagram indicates the order of the play of the cards (positions), not the value of the cards.

3. Ask the person which column contains the chosen card.
4. Pick up the columns with the selected column in the middle.
5. Repeat steps 2, 3, and 4 two more times.
6. Silently count the cards and identify the fourteenth card as the chosen card.

Magic Deal

1. Before the trick, separate a deck of cards into a stack of 15 cards and a stack of the remaining cards.
2. Hand the smaller stack face down to a volunteer.
3. Have another person choose a card from the bigger stack, and show the card to everyone EXCEPT you and the volunteer.
4. Have the chosen card placed on the bottom of the smaller stack held by the volunteer. Make certain the volunteer does not see the card.
5. Have the volunteer do the MAGIC DEAL to identify the card.

    MAGIC DEAL: Place the top card on the table. Place the second card on the bottom of the deck. Place the third card on the table. The next card is placed on the bottom of the deck. Continue until only one card is left.

6. Tell the volunteer to be dramatic. As a card is placed on the table, comment that it *cannot* be the chosen card. As a card is placed on the bottom, comment how it *might* be the card. The audience needs to concentrate, etc.
7. Have the volunteer show the last card to the audience. It will be the chosen card.

Card Flip

1. Place three aces from a deck of cards on a table so that any or all of them are face up.
2. Count the number of aces that are face up. Note whether there is an even or odd number of them.
3. Turn around so you cannot see the cards.
4. Have someone turn the cards over, two at a time.
5. Repeat step 4 any number of times.
6. Have someone cover one of the cards.
7. Turn around and count the number of aces.
8. Predict the covered card:
    If step 2 and step 7 are BOTH EVEN or BOTH ODD, the card is face down.
    If step 2 and step 7 DIFFER, the card is face up.

© Instructional Fair • TS Denison   70   IF2506 Math Magic: Slick Tricks with Numbers

Card Predictor

1. Hand a deck of cards to a volunteer.
2. Have the volunteer shuffle the deck and choose two cards.
3. The volunteer keeps one card and hands the other back for you to see.
4. Explain that the cards have the following values: aces—1, twos through tens—face value, jacks—11, queens—12, and kings—13. The card returned to you will have the same value except that tens and face cards will count as 0.
5. The volunteer looks at his or her card and places it face down on the table.
6. You place the returned card face down to the right of the volunteer's card.
7. Have the volunteer double the value of his or her card, add 2 to the product, and multiply the sum by 5.
8. Unknown to the volunteer, subtract your card's value from 10.
9. Have the volunteer subtract your difference from the last product.
10. Have the volunteer announce the answer.
11. Turn over the two cards—they will show the answer.

Three Card Guess

Have someone choose three cards from a deck, aces through 9, and then:
1. Double the number of one card.
2. Add 5 and multiply by 5.
3. Add the number from one of the two remaining cards.
4. Multiply by 10.
5. Add the number from the third card.
6. Have the person tell you the result.

"Predict" the original numbers by subtracting 250.

Mind Reading Cards

Make the following cards:

1	3	5	7
9	11	13	15
17	19	21	23
25	27	29	31

2	3	6	7
10	11	14	15
18	19	22	23
26	27	30	31

4	5	6	7
12	13	14	15
20	21	22	23
28	29	30	31

8	9	10	11
12	13	14	15
24	25	26	27
28	29	30	31

16	17	18	19
20	21	22	23
24	25	26	27
28	29	30	31

STEPS

1. Have someone think of a number *between* 0 and 32.

2. Show each card and ask if the chosen number is located on the card.

3. For each card containing the chosen number, note the number in the upper left-hand corner.

4. The sum of the upper left-hand numbers will be the chosen number.

Four Cups

1. Set four cups right side up.
2. Place an object in one of the cups. Mentally note whether it is in an EVEN or ODD cup.
3. Turn around.
4. Have someone "pour" the object from the cup to a cup NEXT to it.
5. Repeat step 4 several times. Every time the object is "poured," change EVEN to ODD or ODD to EVEN.
6. When the pouring is finished—and the end result is
    ODD, have the person remove cup 4, and "pour" the object again. You predict cup 2.
    EVEN, have the person remove cup 1, and "pour" the object again. You predict cup 3.

Three Cups

1. Set three cups upside down. Note the starting position of one of the cups (1, 2, or 3). Turn around.
2. Have someone place the object under one of the cups.
3. Have someone interchange the two empty cups.
4. Have someone interchange the cups two at a time stating the LOCATIONS of the cups moved: i.e., 1 and 3, 2 and 3, etc. The object moves with its cup.
5. As you are told the moves, "track" the location of the cup you chose.
6. Repeat step 4 several times.
7. When step 5 is finished, turn around. Predict the object to be under the cup that is at the LOCATION you were "tracking." BE DRAMATIC!

Seasonal Magic Activities

Teacher Notes

The activities in this section present math patterns, computational tricks, modular math, geometric numbers, and shifts in seasonal contexts. "Calendar Squares" could be used any time during the year. The numbers used in "Columbus Day Magic" could be adapted to other dates. Students could be encouraged to create their own geometric shift puzzles based on "April Showers, May Flowers."

Calendar Squares

Activity Students explore square arrays of dates from a calendar.

Comments The "Calendar Squares" activity sheet provided could be replaced by the calendar of any month of the school year.

Any square of nine dates on a calendar can be written as:

A	A + 1	A + 2
A + 7	A + 8	A + 9
A + 14	A + 15	A + 16

Each diagonal has a sum of 3A + 24, (A + A + 8 + A + 16 or A + 2 + A + 8 + A + 14), which equals 3(A + 8) or 3 times the center number.

The sum of all nine numbers is 9A + 72 which equals 9(A + 8) or 9 times the center number.

Other prediction tricks could include:
Given the sum of the square, predict
- the smallest number of the square: sum/9 - 8
- the largest number of the square: sum/9 + 8

Given the sum of either diagonal, predict
- the smallest number: diagonal sum/3 - 8
- the largest number: diagonal sum/3 + 8

Given the sum of the square (the diagonal), predict
- any number of the square by dividing by 9 (or 3) to obtain the middle number (A + 8), then adding or subtracting the appropriate amount based on the values given in the square above.

Columbus Day Magic

Activity Students choose five members from an array, making certain that no two are in the same row or column. The sum of the five is 1492.

Comments The array of numbers is written so that each row (column) entry has certain characteristics that guarantee the desired sum regardless of which entry in the row is used.

The second row entries are each 15 greater than the first row entries. The third row entries are each 14 greater than the second row entries. The fourth row entries are each 13 greater than the third row entries, and the fifth row entries are 9 greater than the fourth row entries.

The second column entries are 104 greater than the corresponding first column entries. The third column entries are 107 greater than the second column entries. The fourth column entries are 102 greater than the third column entries, and the fifth column entries are 99 greater than the fourth column entries.

If the first row, first column entry is called A, then the sum of any five numbers (no two in the same row or column) is 5A + 1,147.

To write an array with a different sum for any five numbers chosen:
1. Subtract 1,492 from the desired sum (ex. 1,997).
2. Divide the difference (1,997 - 1,492 = 505) by 5 (101).
3. Add the quotient to each entry so that no matter which five numbers are chosen, the sum will be five times that quotient greater than 1,492.

The number of ways to obtain the sum from the array:
There are 25 possible first choices. Once the row and column containing the first number are eliminated (9 numbers eliminated), 16 numbers remain for the second choice. Eliminating the row and column containing the second number results in 9 numbers remaining for the third choice. Eliminating the row and column containing the third number leaves 4 numbers for the fourth choice. The elimination of the row and column containing the fourth number, results in 1 number for the fifth choice. The number of sequences to obtain the sum is 25 x 16 x 9 x 4 x 1 = 14,400.

November—Ninth Month?

Activity Students are able to predict a number chosen by other students after following a set of directions.

Comments Assume a three-digit number was chosen $(100h + 10t + u)$. The number can be written as $99h + h + 9t + t + u$ or $99h + 9t + h + t + u$. When the number is divided by 9, the remainder is the same as the remainder of the division of the sum of the digits $(h + t + u)$ and 9—$99h + 9t$ is always divisible by 9.

Assume the tens digit is the digit crossed out. The new dividend is $10h + u$ or $9h + h + u$. Again the remainder is the same as the remainder of the division of the sum of the digits $(h + u)$ by 9. Subtracting the second remainder from the first $[(h + t + u)/9 - (h + u)/9 = t/9]$ is the crossed-out digit t.

If the first remainder is smaller than the second, the sum of the digits of the original number contained an additional 9. Therefore a 9 is added to the first remainder before the second remainder is subtracted.

Example: $357/9 = 39$ R 6 and $(3+5+7)/9$ has a remainder of 6.
Cross out the 5.
$37/9 = 4$ R 1 and $(3+7)/9$ has a remainder of 1.
Subtract remainders: $6 - 1 = 5$, the crossed-out digit.

November Now!

Activity Students follow two sets of directions which always result in a number that is divisible by 11.

Comments TRICK 1: Assume a 4-digit number is used. $1000T + 100h + 10t + u$ is added to $1000u + 100t + 10h + T$. The sum is $1001T + 110h + 110t + 1001u$. Since 1001 and 110 are divisible by 11, the sum is divisible by 11.

TRICK 2: Assume a 3-digit number is used $100h + 10t + u$. Its reverse is $100u + 10t + h$. Regardless of which number is larger, the difference is $99h + 99u$. Since 99 is divisible by 11, the difference is divisible by 11.

Twelve Days of Christmas

Activity Students explore some number patterns based on the song, "The Twelve Days of Christmas."

Comments The twelve days of Christmas are the days from Christmas to the Epiphany, not December 14–25.

The first series of numbers: 1, 3, 6, 10, 15, etc., is the series of triangular numbers. For example,

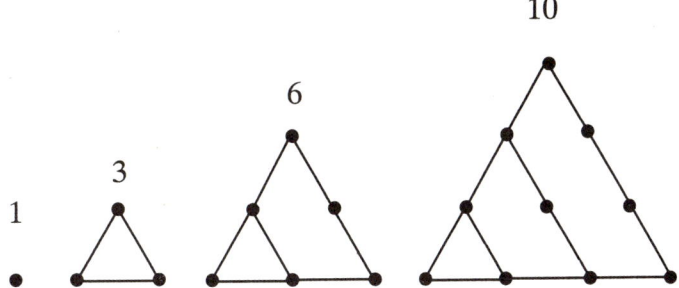

The series of numbers formed by adding the numbers in pairs (first + second, second + third, third + fourth, . . .) is the series of square numbers. See the "Geometric Numbers" activities.

Happy Birthday, Mr. President

Activity Students follow a set of directions to calculate the day of the week of the birthday of a president born since 1900. Each student then finds the day of his or her birth.

Comments The numbers given in the Month Chart depend on the number of days in the birth year used, the number of years the birth year is beyond a multiple of 4 (the leap years), and the number of days in the birth month. The numbers given in the Day Chart are the numbers in the decimal equivalent of $\frac{1}{7}$, 0.142857. They are not in the same order because the number of days in a year is not a multiple of 7. This information is used in the making of a perpetual calendar.

The steps can be used for any date since 1900.

A Mathematical Valentine

Activity Students create a "mathematical heart" by connecting points on a circle using the rule n → 2n.

Comments The activity results in a "heart" generated by computing $2n_{\text{Mod }36}$.

The number system is a modular system of Mod 36 using the whole numbers 0 through 35, much like a clock is the modular system of Mod 12 using the whole numbers 1 through 12.

In this system, a counting method can be used: 27 + 34 is found by starting at 27 and counting 34 clockwise. The answers can also be computed by finding the remainder when the "answer in base 10" is divided by 36: $27 + 34 = 61_{base\ 10}$. $61 \div 36 = 1\ r\ 25$; so $27 + 34 = 25$.

Exploring math concepts of identities and inverses leads to some interesting results. For example:
16 - 16 = 0 and 16 + 20 = 0,
    so -16 and +20 are both additive inverses for 16.
12 x 1 = 12 and 12 x 4 = 12,
    so 1 and 4 are both multiplicative inverses for 12.
Explore the geometric results of using other relations, such as connecting n to 3n or n to 4n.

Shamrock Story

Activity Students model a poem about shamrock leaves to find several possible solutions.

Comments After reading the story once, there appears to be only one solution. When students compare and explain their results, they discover different interpretations.

"Kevin walked alone leaving one leaf on three." can have several interpretations. It could mean that ONLY three leaves are left—one on each of three shamrocks. It could also mean that three shamrocks had only one leaf, while the other shamrocks were left untouched by Kevin. Also the 5 leaves taken by Kary and Katy could have been taken from 2, 3, 4, or 5 different shamrocks. Four possible answers are shown. There are other possible answers. This activity makes a good cooperative group activity. Have students explain their interpretations.

Let 1 = Kary, 2 = Katy, and 3 = Kevin:

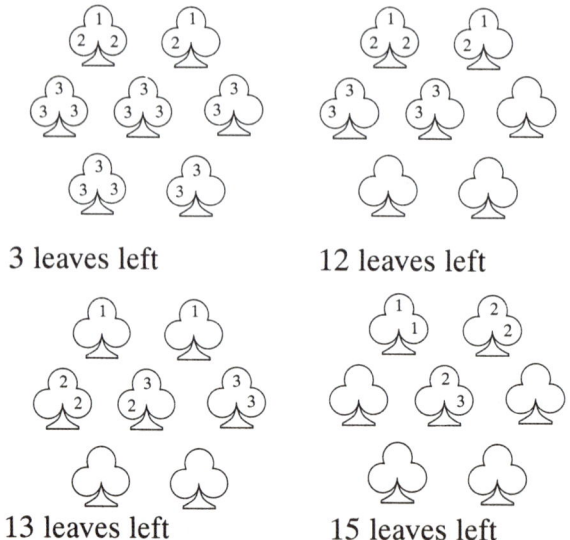

3 leaves left   12 leaves left

13 leaves left   15 leaves left

April Showers, May Flowers

Activity Students cut out three puzzle pieces and rearrange them. The first arrangement shows seven May flowers. The second arrangement shows seven flowers and two raindrops, the April showers.

Comments The trick involves some of the flowers "becoming smaller" and part of a stem becoming part of a flower in order to have the raindrops "appear." A similar puzzle involving length is shown below.

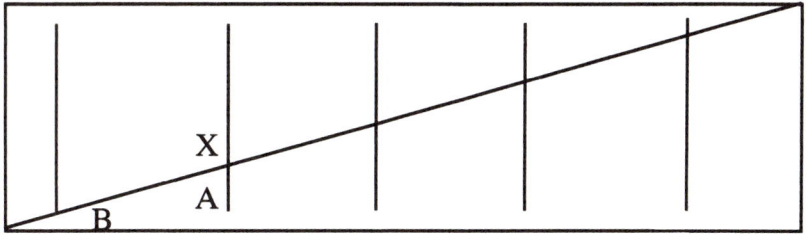

Cut along the diagonal and shift X from point A to point B. It "appears" that five lines become six lines. Note that the sum of all lengths is the same before and after the shift.

Name _____

Calendar Squares

Draw a box around any nine dates forming a square.

SUN.	MON.	TUES.	WED.	THURS.	FRI.	SAT.
	1	2	3	4	5	6
7	8	9	10	11	12	13
14	15	16	17	18	19	20
21	22	23	24	25	26	27
28	29	30	31			

1. Find the sum of one of the diagonals. _____

2. Find the sum of the other diagonal. _____

3. How does a diagonal sum compare with the center number? _____

4. Find the sum of all nine numbers. _____

5. How does the sum of all nine compare to the center number? _____

Draw a box around another set of nine dates forming a square.

Repeat steps 1 through 5.   1. _____   2. _____

3. _____   4. _____

5. _____

Given the sum of the nine numbers in a square, how could one predict the numbers in the square?

_____

© Instructional Fair • TS Denison   80   IF2506 Math Magic: Slick Tricks with Numbers

Name _____

Columbus Day Magic

Select any number from the array of numbers below. Now select any other number not in the same row or column as the first number. Do the same for five numbers so that no two numbers are from the same row or column.

Find the sum of the five numbers. _____
How does the sum relate to the title of the activity?

_____

69	173	270	372	471
84	188	285	387	486
98	202	299	401	500
111	215	312	414	513
120	224	321	423	522

Choose another five numbers. _____ What is the sum? _____

To discover how many ways you can obtain this sum, answer the following questions.

1. How many numbers are there for the first choice? _____

2. If you cross out the column and row containing the first choice, how many numbers remain for the second choice? _____

3. Again, cross out the column and row containing the second choice; how many numbers remain for the third choice? _____

4. Cross out the column and row. How many numbers remain for the fourth choice? _____

5. Cross out the column and row. How many numbers remain for the fifth choice? _____
   Multiply the answers together _____ x _____ x _____ x _____ x _____ = _____
   The product is the number of ways you can get the sum!

November—Ninth Month?

November comes from the Latin *nove* or nine. In the early Roman Republican Calendar, September, October, November, and December were the seventh, eighth, ninth, and tenth months of the year respectively. In England before 1752, March 25 was the beginning of a new year. The acceptance of the Gregorian Calendar made January 1 the first day of the year. November was no longer the ninth month.

Amaze your family and friends (at your Thanksgiving gathering) with this trick based on the number 9.

Have a friend . . .

1. choose any number of three or more digits.

2. divide the number by 9.

3. tell you the remainder.

4. erase any one non-zero digit of the dividend.

5. divide the new number by 9.

6. tell you the new remainder.

You:

7. announce the crossed-out digit by subtracting the second remainder from the first remainder.

Note: If the second remainder is larger than the first, add 9 to the first remainder before subtracting.

Name _____

November Now!

November has been the eleventh month of the year since the Gregorian Calendar made January 1 the accepted first day of the year. Prior to the acceptance of the Gregorian Calendar, November was the ninth month. Its name is from the Latin for nine. Even England had the new year starting in March prior to 1752.

Here are two tricks with 11 that might amaze you and your friends.

TRICK 1:

1. Choose any number that has an **even** number of digits (2-digit, 4-digit, 6-digit, etc.)

2. Reverse the digits of the chosen number.

3. Add the number and its reverse.

4. Divide the sum by 11.

5. Is there a remainder?

6. Try other numbers. Do you ever get a remainder?

TRICK 2:

1. Choose any multi-digit number that has an **odd** number of digits (3-digit, 5-digit, 7-digit, etc.).

2. Reverse the digits of the chosen number.

3. Subtract the smaller number from the larger.

4. Divide the difference by 11.

5. Is there a remainder?

6. Try other numbers. Do you ever get a remainder?

Name _____

Twelve Days of Christmas

According to the popular song "The Twelve Days of Christmas," on the twelfth day of Christmas the true love gave:

Twelve pipers piping,
Eleven drummers drumming,
Ten lords a leaping,
Nine ladies dancing,
Eight maids a milking,
Seven swans a swimming,
Six geese a laying,
Five golden rings,
Four calling birds,
Three French hens,
Two turtle doves, and
A partridge in a pear tree.

1. How many gifts were given on the

   twelfth day? _____   eleventh day? _____   tenth day? _____

   ninth day? _____   eighth day? _____   seventh day? _____

   sixth day? _____   fifth day? _____   fourth day? _____

   third day? _____   second day? _____   first day? _____

2. List the numbers from smallest to largest. _____

3. Make another series by adding the numbers in pairs (first + second, second + third, third + fourth, . . .).
   _____

4. Describe the second series of numbers. _____

5. What is the total number of gifts given during the twelve days of Christmas? _____

© Instructional Fair • TS Denison   84   IF2506 MATH MAGIC: Slick Tricks with Numbers

Name _____

Happy Birthday, Mr. President!

Originally, President's Day was to celebrate the February birthdays of George Washington, February 22, 1732 (February 11, 1731, old calendar) and Abraham Lincoln, February 12, 1809. Now, the day is set aside to recognize all presidents of the United States.

President John F. Kennedy was the first president to be born in the twentieth century. The presidents' birthdays of the 1900s are

- John F. Kennedy, May 29, 1917
- Lyndon Johnson, August 27, 1908
- Richard Nixon, January 9, 1913
- Gerald Ford, July 14, 1913
- Jimmy Carter, October 1, 1924
- Ronald Reagan, February 6, 1911
- George Bush, June 12, 1924
- William Clinton, August 19, 1946

Follow the steps to compute the day of the week on which one of the presidents above was born.

Month Chart

Month	Number
January	1 (0 leap yr.)
February	4 (3 leap yr.)
March	4
April	0
May	2
June	5
July	0
August	3
September	6
October	1
November	4
December	6

1. Write the last two digits of the year. _____
2. Divide by 4 (drop the remainder). _____
3. Write the month number from the chart. _____
4. Write the date of the month. _____
5. Add the answers from 1–4. _____
6. Divide by 7 (round two decimal places). _____
7. Write the digit in the tenths place. _____
8. Find the day on the Day Chart. _____

Day Chart

Sunday 1	Monday 2
Tuesday 4	Wednesday 5
Thursday 7	Friday 8
Saturday 0	

Find the day of the week of your birthday.

A Mathematical Valentine

Name _____

Use a red pencil and a ruler to connect the numbers using the relationship: n connects to 2n. For example: 1→2, 2→4, 3→6, . . . , 17→34, 18→0, 19→2, etc.

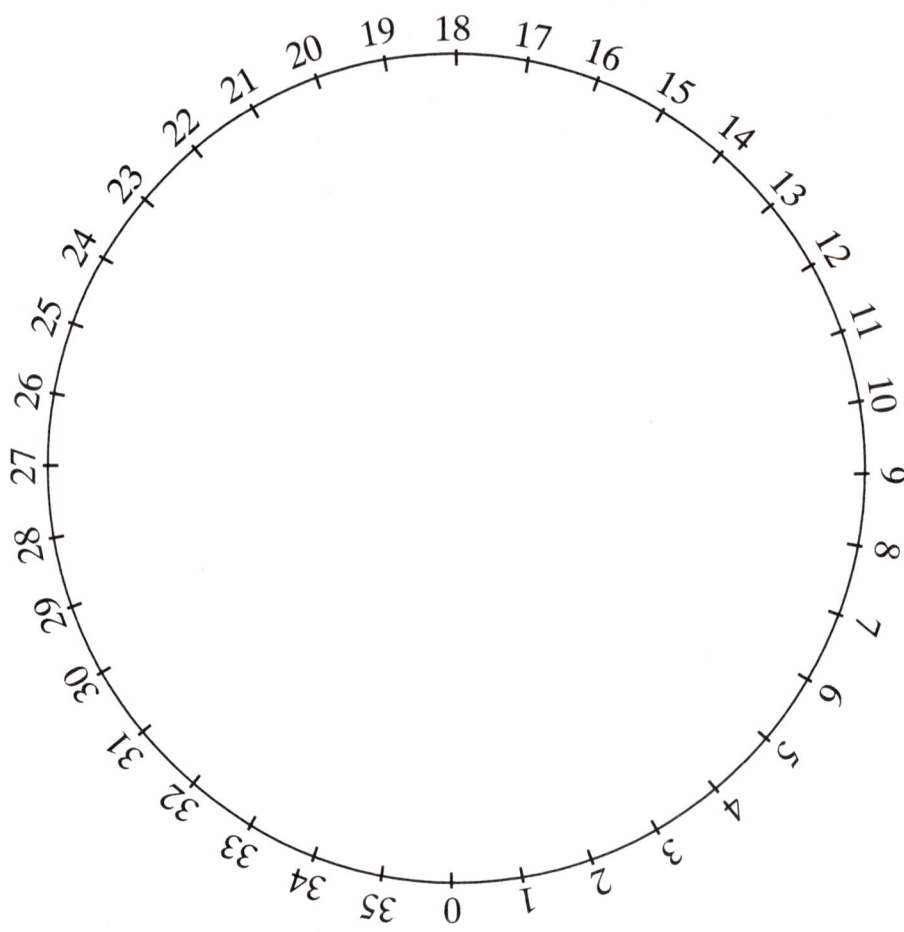

Assume a number system only uses the whole numbers 0 through 35 like the diagram above. Compute the following:

1. 35 + 1 = _____   2. 27 + 34 = _____   3. 5 - 8 = _____

4. 18 - 29 = _____   5. 6 x 6 = _____   6. 2 x 35 = _____

7. 27 - 27 = _____ and 27 + 9 = _____, so -27 and 9 are both additive inverses for 27.

8. 6 x 1 = _____ and 6 x 7 = _____, so 1 and 7 are multiplicative identities for 6.

9. Explain how you computed an answer that in base 10 would be greater than 35.

_____

_____

© Instructional Fair • TS Denison   86   IF2506 Math Magic: Slick Tricks with Numbers

Name _____

Shamrock Story

Use the "shamrock fields" below to model the poem. Can you find more than one solution?

Three wee leprechauns, Kary, Katy, and Kevin
Went into a field that had shamrocks numbering seven.
Mischievous Kary took two leaves along the way,
While for cute Katy taking three leaves made her day.
Kevin walked alone leaving one leaf on three.
How many leaves remaining might there be?

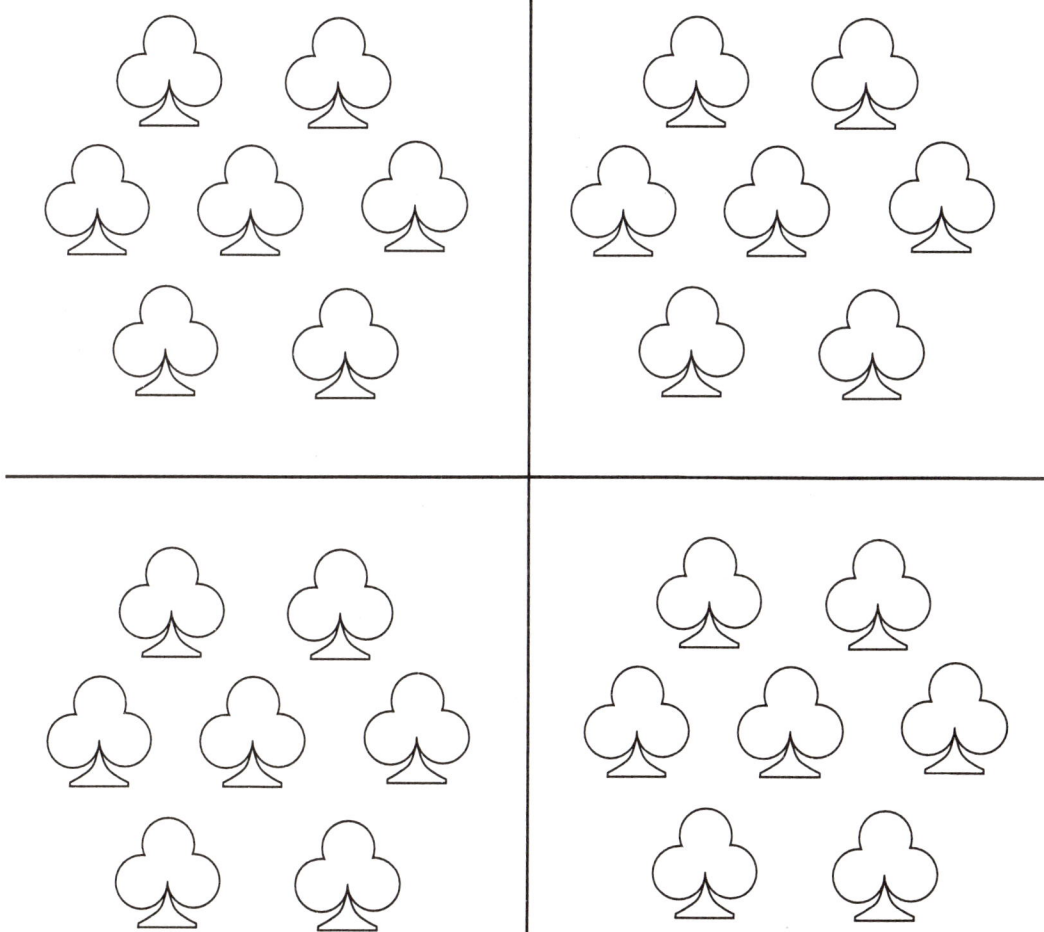

Name _____

April Showers, May Flowers

It is often said that "April showers bring May flowers." Cut out the rectangle containing the May flowers. Cut along the solid center line. Cut along the dashed line to separate pieces 1 and 2. Interchange pieces 1 and 2. Can you find the April showers?

Answer Key

Switch-a-Roo — **Page 7**

2,418 and 2,418	1,008 and 1,008
806 and 806	756 and 756
1,209 and 1,209	1,512 and 1,512
2,016 and 2,016	504 and 504

For each pair of problems, the product of the tens digits of the factors equals the product of the ones digits.

Magical Nines — **Page 8**

1. 7,992
   79,992
   799,992
   7,999,992
   79,999,992

Number of digits in product is one more than number of 9s in the factor. First digit is 7, last digit is 2, other digits are 9s.

2.

2,277		23	2,277
3,366	34	34	3,366
4,455	45	45	4,455
5,544	56	56	5,544
6,633	67	67	6,633
7,722	78	78	7,722
8,811	89	89	8,811

100n - n. Also the first two digits plus the last two digits equal 99.

3. 18,218
   27,327
   36,436
   45,545
   54,654
   63,763
   72,872
   81,981

First 2 digits n x 9, middle digit n, last 2 digits n x 9

Could You Repeat That? — **Page 9**

$0.33\overline{3}$	$0.11\overline{1}$	$0.09\overline{09}$
$0.66\overline{6}$	$0.22\overline{2}$	$0.18\overline{18}$
$0.99\overline{9}$ or 1	$0.33\overline{3}$	$0.27\overline{27}$
	$0.44\overline{4}$	$0.36\overline{36}$
$0.\overline{142857}$	$0.55\overline{5}$	$0.45\overline{45}$
$0.\overline{284714}$	$0.66\overline{6}$	$0.54\overline{54}$
$0.\overline{428571}$	$0.77\overline{7}$	$0.63\overline{63}$
$0.\overline{571428}$	$0.88\overline{8}$	$0.72\overline{72}$
$0.\overline{714285}$	$0.99\overline{9}$ or 1	$0.81\overline{81}$
$0.\overline{857142}$		$0.90\overline{90}$
1		$0.99\overline{99}$ or 1

1. Numerator times $0.\overline{3}$
   Numerator times $0.\overline{1}$
2. Numerator times $0.\overline{09}$
3. The digits 142,857 occur in the same order starting at different places
4. Example: $1 = \frac{1}{3} + \frac{2}{3} = 0.3\overline{3} + 0.6\overline{6} = 0.9\overline{9}$

Are All Things Equal? — **Page 10**

			4			2	
		$\frac{1}{4}$	5		4		$\frac{3}{4}$
	6	$\frac{1}{5}$	6		5	4	$\frac{4}{5}$
$\frac{1}{6}$	7	$\frac{1}{6}$	7		$\frac{5}{6}$	5	$\frac{5}{6}$

		$\frac{3}{4}$				4	
	$\frac{1}{5}$	$\frac{4}{5}$			$\frac{1}{4}$	5	
$\frac{5}{6}$	$\frac{1}{6}$	$\frac{5}{6}$		6	$\frac{1}{5}$	6	
$\frac{1}{7}$	$\frac{6}{7}$	$\frac{1}{7}$	$\frac{6}{7}$	$\frac{1}{6}$	7	$\frac{1}{6}$	7

$1\frac{1}{7} \text{ x } 8 = 1\frac{1}{7} + 8 = 9\frac{1}{7}$

$6\frac{1}{8} \div \frac{7}{8} = 6\frac{1}{8} + \frac{7}{8} = 7$

$6 \text{ x } \frac{6}{7} = 6 - \frac{6}{7} = 5\frac{1}{7}$

$9\frac{1}{7} \div 8 = 9\frac{1}{7} - 8 = 1\frac{1}{7}$

Geometric Numbers — **Page 11**

1. 1, 3, 6, 10, 15, 21, 28, 36, 45, 55
2. 2, 3, 4, 5, 6, 7, 8, 9, 10
   Consecutive integers ≥ 2 or 1 + 1n
   (n is the number of the term)
3. 4, 9, 16, 25, 36, 49, 64, 81, 100
   Squares ≥ 4
4. 1, 4, 9, 16, 25, 36, 49, 64, 81, 100
5. 3, 5, 7, 9, 11, 13, 15, 17, 19
   Consecutive odds ≥ 3 or 1 + 2n
6. 1, 5, 12, 22, 35, 51, 70, 92, 117, 145
7. 4, 7, 10, 13, 16, 19, 22, 25, 28; 1 + 3n
8. 1, 6, 15, 28, 45, 66, 91, 120, 153, 190
9. 5, 9, 13, 17, 21, 25, 29, 33, 37; 1 + 4n

"Sum" Fun with Fibonacci—I — **Page 13**

Answers will vary.
3. and 4. The answers will be equal.

"Sum" Fun with Fibonacci—II — **Page 14**

Answers will vary.

Under 233: 609	Under 1,597: 4,180
Under 377: 986	Under 6,765: 17,710

"Sum" Products — **Page 15**

1.-6. Answers will vary.
7. The products differ by 1.
8.-11. Answers will vary.
12. The products differ by 1.

Was Fibonacci a Square? — **Page 16**

$1^2 = 1$	$21^2 = 441$
$2^2 = 4$	$34^2 = 1,156$
$3^2 = 9$	$55^2 = 3,025$
$5^2 = 25$	$89^2 = 7,921$
$8^2 = 64$	$144^2 = 20,736$
$13^2 = 169$	$233^2 = 54,289$

Squares: 1; 4; 9; 25; 64; 169; 441; 1,156; 3,025; 7,921; 20,736; 54,289

Sums: 5; 13; 34; 89; 233; 610; 1,597; 4,181; 10,946; 28,657; 75,025
Differences of sums: 8; 21; 55; 144; 377; 987; 2,584; 6,765; 17,711; 46,368
Sums: odd terms of Fibonacci series
Differences: even terms of Fibonacci series

Chet: The Sum Checker — Page 22
1. 2
   7
   <u>6</u>
   15→6
   15→6 OK
2. 8
   6
   <u>8</u>
   22→4
   21→3 No
3. 4
   3
   <u>0</u>
   7
   24→6 No
4. 0
   1
   <u>4</u>
   5
   14→5 OK

5. 3
   <u>8</u>
   −5 + 9 = 4 OK
   13→4 OK
6. 3
   <u>7</u>
   −4 + 9 = 5
   14→5 OK
7. 4
   <u>6</u>
   −2 + 9 = 7
   24→6 No
8. 0
   <u>4</u>
   −4 + 9 = 5
   5 OK

Chet: The Product Checker — Page 23
1. 2
   x4
   8
   26 →8 OK
2. 8
   x6
   48 →3
   20 →2 No
3. 4
   x3
   12 →3
   30→3 OK
4. 0
   x1
   0
   17→8 No

5. 3
   x3
   9 →0
   36→0 OK
6. 6
   x3
   18 →0
   26→8 No
7. 2
   x7
   14 →5
   23→5 OK
8. 2
   x3
   6
   15→6 OK

Multiplication Madness — Page 24
29 x 38
~~14 x 76~~
7 x 152
3 x 304
1 x <u>608</u>
1,102

~~44 x 50~~
~~22 x 100~~
11 x 200
5 x 400
~~2 x 800~~
1 x <u>1,600</u>
2,200

65 x 12
~~32 x 24~~
~~16 x 48~~
~~8 x 96~~
~~4 x 192~~
~~2 x 384~~
1 x <u>768</u>
780

~~78 x 61~~
39 x 122
19 x 244
9 x 488
~~4 x 976~~
~~2 x 1,952~~
1 x <u>3,904</u>
4,758

~~80 x 90~~
~~40 x 180~~
~~20 x 360~~
~~10 x 720~~
5 x 1,440
~~2 x 2,880~~
1 x <u>5,760</u>
7,200

Multiplication Magic — Page 25
1. 4 x 5 = 20
2. 3,025
3. 4,225
4. 5,625
5. 7,225
6. 9,025
7. 13,225
8. 15,625
9. 18,225
10. 21,025
42,025

Win Some, Lose Some Multiplication — Page 26
1. 1   1
2. 2   2   396
3. 3   3   9   1591
4. 90 + 4   90 − 4   16   8,084
5. 80 − 5   80 + 5   6,400 − 25   6,375
6. 40 + 6   40 − 6   1,600 − 36   1,564
7. 50 + 7   50 − 7   2,500 − 49   2,451

Pascal's Triangle — Page 27
1. 1 6 15 20 15 6 1
   1 7 21 35 35 21 7 1
2. Each number is the sum of two numbers "above."
3. 1 4 6 4 1
4. 1 5 10 10 5 1
5. $1x^6+6x^5y^1+15x^4y^2+20x^3y^3+15x^2y^4+6xy^5+1y^6$

Integer Trees — Page 28

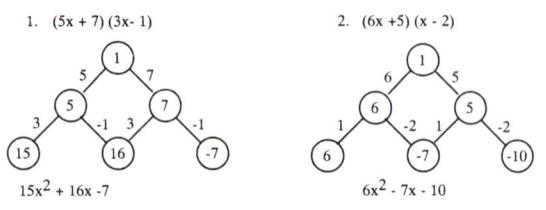

$15x^2 + 16x - 7$ $6x^2 - 7x - 10$

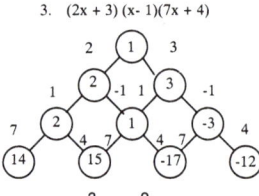

$14x^3 + 15x^2 - 17x - 12$

Back to the Beginning — Page 38
BTTB 1.–4. Answers will vary.
Last number equals beginning number.

Age Predictor — Page 40
Answer is birth month followed by age.

Birthday Predictor — Page 40
Answer is month, day, and year of birthday.

Pick a Number—Any Number — Page 41
1.–4. Answers will vary.
9s or 0s

Calculator Grid — Page 41
1.–2., 4. Answers will vary.
3. Always a multiple of 198.
5. 198

Double Vision — Page 42
1.–4. Answers will vary.
5. The six-digit number will be the original number repeated.

Multi-vision — Page 42
1.–2. Answers will vary.
3. Original number repeated 6 times.
4. Multi-vision occurs because the number repeats.

Caught in the Middle   Page 43
"Sums" are the middle digits.

Digit Predictor   Page 43
The result is the "crossed out" digit.

Sum Predictor   Page 44
The result is always 1,998 more than the original number. "Subtract 2 and place it in front."

Human Calculator   Page 45
Answers will vary. The last two digits will be the sum of the units of the five numbers chosen, and the first two digits will be the difference of 40 and the sum.

Fantastic Fractions   Page 46
1.-6. Answers will vary.
7.-8. Same as 1. and 2.

High Noon   Page 47
Final number: 12

A Mouse in the House   Page 48
Max ends up in the laundry room.

Magic Squares   Page 56
One possible solution: Sum: 15

8	1	6
3	5	7
4	9	2

Sum: 68

28	2	24	14
22	16	26	4
10	20	6	32
8	30	12	18

Fraction Magic Squares   Page 57
One possible solution: Sum: 4 ¼

7/4	1/8	3/2	7/8
11/8	1	13/8	1/4
5/8	5/4	3/8	2
1/2	15/8	3/4	9/8

Sum: 3⅖

7/5	1/10	6/5	7/10
11/10	4/5	1 3/10	1/5
1/2	1	3/10	1 3/5
2/5	3/2	3/5	9/10

Decimal Magic Squares   Page 58

Sum: 2.5

.9	.25	.8	.55
.75	.6	.85	.3
.45	.7	.35	1
.4	.95	.5	.65

Sum: 1.7

.7	.05	.6	.35
.55	.4	.65	.1
.25	.5	.15	.8
.2	.75	.3	.45

Equation Magic Square   Page 59

Sum: 130

34	48	2	16	30
46	10	14	28	32
8	12	26	40	44
20	24	38	42	6
22	36	50	4	18

A Magical Magic Square   Page 59

		3c C	
	11 F		
9 I		6 K	12 L
	14 N	15 O	

1. 34
2. 34
3. 34
4. 34
5. 34
6. 748
7. 748

Magic Square Triangle Page 59
A 3-4-5 triangle is a right triangle.

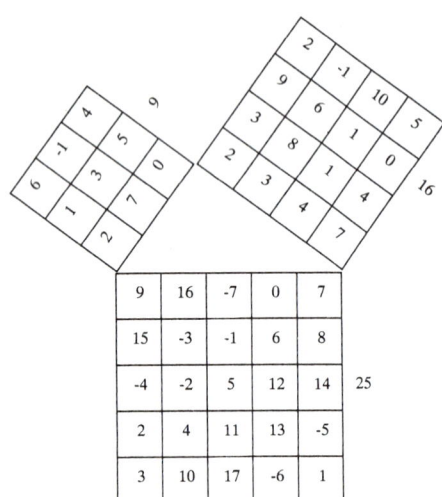

Magic Triangles Page 62
Answers may vary. Possible solutions:

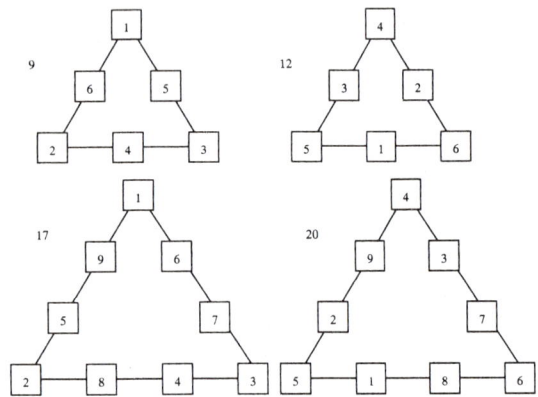

The largest magic sum of a magic triangle using digits 1 through 9 is 23.

Magic Circles Page 63
Answers may vary. Possible solutions:

Use the digits 1 through 7.
Magic Sum = 12

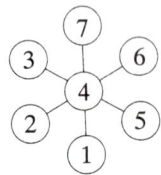

Use the odds 1 through 13.
Magic Sum = 21

Use the first 9 positive multiples of 3.
Magic Sum = 45

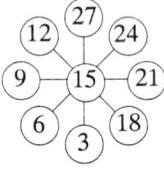

Use the first 11 positive multiples of 5.
Magic Sum = 90

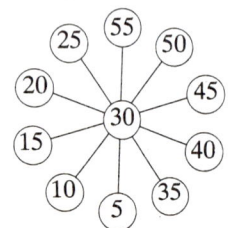

Calendar Squares Page 80
1.-2. Answers will vary.
3. Three times the center number.
4. Answers will vary.
5. Nine times the center number.
   Sum/9 will be the center number. Subtract 8, 7, 6, 1, and add 1, 6, 7, and 8 to find the other 8 numbers.

Columbus Day Magic Page 81
1492
"In 1492, Columbus sailed the ocean blue."
Numbers chosen will vary, but answer will always be 1492.
1. 25
2. 16
3. 9
4. 4
5. 1
25 x 16 x 9 x 4 x 1 = 14,400

November—Ninth Month? Page 82
Answers will vary. Result is always the crossed-out digit.

November Now! Page 83
Answers will vary. Results are always divisible by 11.

Twelve Days of Christmas Page 84
1. 78   66   55
   45   36   28
   21   15   10
   6   3   1
2. 1, 3, 6, 10, 15, 21, 28, 36, 45, 55, 66, 78
3. 4, 9, 16, 25, 36, 49, 64, 81, 100, 121, 144
4. Squares of 2 and greater
5. 364

Happy Birthday, Mr. President! Page 85
Kennedy—Tuesday   Johnson—Thursday
Nixon—Thursday   Ford—Monday
Carter—Wednesday   Reagan—Monday
Bush—Thursday   Clinton—Monday
Answers will vary.

A Mathematical Valentine Page 86
1. 0   2. 25   3. 33
4. 25   5. 0   6. 34
7. 0   0
8. 6   6
9. Possible answer, divide by 36, use the remainder.

April Showers, May Flowers Page 88

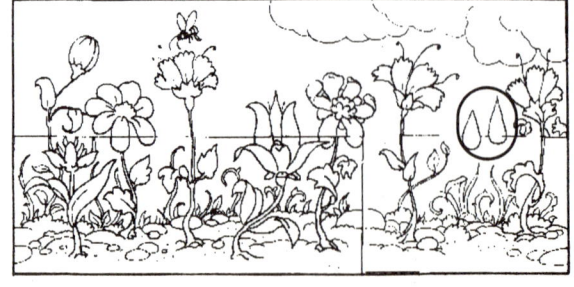